PUBLICATIONS

FRACTALS
A USER'S GUIDE
FOR THE NATURAL
SCIENCES

HAROLD M. HASTINGS
AND GEORGE SUGIHARA

Fractals

Fractals

A User's Guide for
the Natural Sciences

HAROLD M. HASTINGS

Hofstra University,
Hempstead, New York

and

GEORGE SUGIHARA

Scripps Institution of Oceanography,
La Jolla, California

Oxford New York Tokyo
OXFORD UNIVERSITY PRESS

Oxford University Press, Walton Street, Oxford OX2 6DP
Oxford   New York   Toronto
Delhi   Bombay   Calcutta   Madras   Karachi
Kuala Lumpur   Singapore   Hong Kong   Tokyo
Nairobi   Dar es Salaam   Cape Town
Melbourne   Auckland   Madrid
and associated companies in
Berlin   Ibadan

Oxford is a trade mark of Oxford University Press

Published in the United States
by Oxford University Press Inc., New York

© Harold M. Hastings and George Sugihara, 1993

First published 1993
Reprinted 1994

A catalogue record for this book is available from the British Library

Library of Congress Cataloging in Publication Data
Hastings, Harold M., 1946–
Fractals : a user's guide for the natural sciences / Harold M.
Hastings and George Sugihara.
Includes bibliographical references.
1. Fractals. I. Sugihara, G. II. Title.
QA614.86.H37 1993 514'.74—dc20   93-12661

ISBN 0 19 854598 3 (Hbk)
ISBN 0 19 854597 5 (Pbk)

Printed in Great Britain on acid-free paper by
Redwood Books Ltd., Trowbridge, Wilts.

To
Gretchen
and
Joan

Acknowledgements

We wish to thank Professor Robert M. May for his constant encouragement and helpful comments, and the staff of Oxford University Press for guiding this project to completion. The section on the geometry of pancreatic islets (8.1) was co-authored by Dr Bruce Schneider, whose help we gratefully acknowledge. We wish to thank Professor Paul Ehrlich for helpful correspondence on population dynamics and Professor John McGowan for bringing the Scripps pier ocean temperatures to our attention. We also wish to thank Diana Beaudette for help with calculations of population time series, Dennis Grace for helpful discussions, and Richard Penner for assistance with programming. Micheline Schreiber helped with computations of the fractal dimension of pancreatic islets and David Troyan helped with earthquake calculations. Finally, we thank many students for helping us learn fractals through teaching.

Contents

Part III The bridge to applications

Part IV Case studies

Part V The toolbox

A Brief Chronology of Fractals

The monsters (complex, irregular objects)

1872: The Cantor set.

1875: Weierstrass's continuous
nowhere-differentiable curve
(du Bois).

1906: Brownian motion (Perrin).
The Koch snowflake (von Koch).

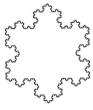

Scaling behaviour

1919: The Hausdorff dimension of complex geometric objects.

1951: Hurst's law for scaling behaviour of Nile river discharges.

1956: The Gutenberg–Richter law for the distribution of earthquake magnitudes.

1961: Richardson's scaling laws in the measurement of complex natural curves such as coastlines.

1963: The Stommel diagram describes spatial and temporal scales for ocean dynamics in space and time.

1968–9: Mandelbrot, Van Ness, and Wallis extend Hurst's work in hydrology.

The science of fractals (except for dynamics)

1975: Mandelbrot coins the word 'fractals'.

1977: Mandelbrot's *Fractals: Form, Chance and Dimension*.

1980: The Weierstrass–Mandelbrot fractal function (Berry and Lewis)—the natural geometry of the Weierstrass monster of 1875.

1982: Fractal models applied to ecology (Hastings *et al.*) and cloud patterns (Lovejoy). Mandelbrot's revised monograph *The Fractal Geometry of Nature*.

1986: Iterated function systems (Barnsley *et al.*).

Fractals and dynamics

1981: Witten and Sander introduce
diffusion-limited aggregation.

1983: Hentschel and Procaccia relate
fractals and strange attractors.

1984: Wolfram's dynamics of cellular
automata.

1987: Bak, Tang, and Weisenfeld's self-organized criticality.

1991: Schertzer and Lovejoy's book on fractals and multifractals in geophysics.

The present

1991: 400 publications.

1992: Fractals applied to answer open questions ranging from the origin and structure of the universe to the distribution of earthquakes.

Part I

Introduction

Euclidean geometry has shaped much of the way natural forms are viewed in science and mathematics, and even in art; and seems to be writ deeply.in the human psyche. Motivated by our basic desire to find simplicity and order in nature, Euclidean ideals are often held out as approximations or caricatures of natural forms that may be essentially complex and irregular. Thus, the planets are roughly spheres, elm leaves are ellipses, and spruce trees are roughly cone-shaped. That is to say, we achieve simplicity by filtering out the complexity and uniqueness of natural forms and identifying their essence with the class of shapes which can be rendered by protractors, conic sections and French curves.

Whether nature is 'essentially' complex (that is, irregular and random) or 'essentially' simple (that is, Euclidean and ordered) is in some sense an artificial dichotomy. Piet Mondrian's geometric forms and Jackson Pollack's random patterns both capture important parts of nature. Can an appropriate geometry combine the complexity of Pollack's patterns with the simplicity of Mondrian's descriptions?

We believe that fractals, and in particular random fractals, may provide such a bridge. To quote the founder of this field, Benoit Mandelbrot (1989): 'Fractals provide a workable new middle ground between the excessive geometric order of Euclid and the geometric chaos of roughness and fragmentation.'

Self-similarity and measurement

The key idea in fractal geometry is *self-similarity*. An object is self-similar if it can be decomposed into smaller copies of itself. Thus self-similarity is the property in which the structure of the whole is contained in its parts. For example, recall that line segments, squares, and cubes are measured by dividing them into similar small units, for example, centimeters, square centimeters, and cubic centimeters. These smaller units are similar to the whole, and share the same units of measurement. That is to say, the basic building blocks (straight lines, squares and cubes) can be summed to produce the 'measure' of the whole.

In the formalism of calculus, highly magnified views of a smooth curve

The Hofstra University Pineum. Henry Moore's *Upright Motive No. 9*, 1979, in the foreground, is an idealized, artistic vision of the human form built from smooth curves and surfaces. Although Euclidean geometry is also evident in the patio in the foreground and building in the background, it does not capture the order inherent in the complex forms of pine trees throughout the photograph. A new geometry—fractal geometry—is needed to provide useful, compact descriptions of such natural forms. (Reproduced by kind permission of the Henry Moore Foundation.)

can be viewed in the limit as a polygonal arc with infinitesimally small sides. Thus, in the limit, straight line segments are the basic building blocks of smooth curves, and the length (measure) of a curve is just the sum of the lengths of infinitesimally small line segments. Indeed, most regular geometric objects can be measured with the same basic building blocks: lines, squares, and cubes.

On the other hand, highly magnified view of natural forms such as coastlines, vegetation patches, and graphs of population fluctuations do not appear smooth and may not reduce to the usual building blocks. Indeed, upon magnification, they will frequently display the same irregularity on the smallest scales as is present in the large. Because of this, such complex objects, called *fractals*, cannot in fact be rigorously measured by the basic building blocks of Euclidean geometry. A fractal curve does not have a 'length' as we commonly know it. Indeed, the basic building block of a self-similar irregular form is an infinitesimally small copy of itself; and its measure comes from enumerating these small irregular building blocks.

Barnsley's 'trademark'—the black spleenwort fern (c. M. F. Barnsley, *Fractals Everywhere*, Academic Press, 1988, Fig. 3.10.8(b), reprinted with permission). This complex pattern can be generated relatively simply through the use of fractals, using a six-parameter iterated function system. It is, however, generally quite difficult to find such descriptions.

While the basic building blocks of Euclidean geometry are simple, the mathematical apparatus required to mimic a complex form can be obtuse. For example, a Euclidean description of the black spleenwort fern might involve a polynomial with thousands of fitted parameters. One is basically pushing a simple basic building block (a straight line) into a complex shape with a complicated function. In contrast, the complexity in fractal geometry comes from the building blocks, and the process which generates the larger pattern is relatively simple. The simple process, recursion, essentially involves echoing a simple rule over and over again, giving rise to self-similar geometry. Indeed the assembly of fractals from their natural building blocks, smaller self-similar fractals, appears to capture some essential aspect of the growth and branching rules of nature itself.

Thus fractal and Euclidean geometry are conjugate approaches to the geometry of natural forms. Fractal geometry builds complex objects by applying simple processes to complex building blocks; Euclidean geometry uses simpler building blocks but frequently requires complex processes. As we shall see, fractal geometry, by emphasizing a unique relationship between a form and its building blocks, seems to fit nature remarkably well.

Mandelbrot's first book (1975, 1977) demonstrated the potential of fractals

To provide a unified setting for the study of ubiquitous irregular objects, regarded as 'mathematical monsters' in the late nineteenth and early twentieth centuries. Since Mandelbrot's (1982) book, *The fractal geometry of nature*, the field of fractals has exploded. The 1991 *Science Citation Index* listed over 400 papers with the words 'fractals' or 'power laws' (the algebraic analogue of fractals) in their title. They span fields ranging from physical geometry (the surface texture of sea beds, the structure of continental faults and, in the spring of 1992, a lively debate about the distribution of intervals between earthquakes) to ecology (fungal structures, the power law relationship between the area of a quadrat and the number of species it contains, the structure of peat systems). Cosmology (the structure of star clusters and galaxies and, in May 1992, questions about the big bang and the origin of the universe), and developmental biology (lung branching patterns, heart rhythms, the structure of neurons) are also represented.

Stimulated by the explosive growth in the science of fractal geometry, we have written this book as both a logically developed text and a handbook for natural scientists interested in applications of fractals—in short, a book on 'fractals for users'. We have aimed to make the book self-contained: most chapters assume only modest mathematical background.

The first chapter introduces and motivates fractals and their algebraic analogues, formulae of the form $y = ax^b$ called power laws, as geometric and algebraic images of nature. Everyone is familiar with the power laws of Euclidean geometry: formulae for the area of a square $A = s^2$ and the volume of a cube $V = s^3$. We shall motivate and explain power laws in nature: classical power laws such as allometric relationships connecting the size and metabolism of an animal, relationships between the area of a quadrat and the number of species it contains (the species–area law), and the Gutenberg–Richter law relating the number and magnitude of earthquakes, as well as several new power laws discovered through fractal geometry.

Part II builds upon these ideas to develop *the mathematics of random fractals*. Chapter 2 introduces the basic foundations: self-similarity, self-affinity, scaling and Hausdorff dimensions, and the mathematics of power laws. The scaling and Hausdorff dimensions are exponents in power laws for measuring the complex objects of nature, just as the familiar topological dimension in the exponent in the formulae $A = s^2$ and $V = s^3$; see also McGuire's (1991) beautifully illustrated essay on fractals.

The next three chapters develop practical techniques for computing the dimension and other scaling exponents associated with fractal patterns and time series. In particular, Chapter 3 develops a wide variety of simple and useful alternatives to the Hausdorff and scaling dimensions. Although all of these alternative dimensions measure essentially the same thing—how to build a complex object simply by using itself as the building block, applications will frequently dictate the choice of one particular technique. Chapter 4 describes similar approaches for the characterization of complex time series. The last of these chapters, Chapter 5, describes Fourier transform techniques and their use in computing fractal exponents. Chapters 2–5 gather together material from many sources, ranging from nineteenth-century examples, largely following the explanations of Mandelbrot (1977, 1982) to papers written in the 1980s and 1990s. Although much of this material is 'well known' to those active in the field, it is not readily accessible to prospective users working in other fields, or present-day students in the natural sciences. We aim in these chapters to steer a clear path between the excessive formality of many papers in pure mathematics and the excessive informality of many books on fractals, in keeping with our goal of a 'middle ground' book for students and researchers in the natural sciences.

The next four chapters form a bridge from the mathematics to two case studies: fractal patterns in vegetative ecosystems, and the new study of the persistence and extinction of small populations. Chapter 6 begins with an introduction to statistical techniques used to compute fractal exponents from experimental data, namely log transforms and linear regression. It also describes tests for fractal behaviour and the role of simple simulation techniques in determining the validity and confidence limits of fractal descriptions. Since a thorough test of the validity of fractal models in a particular application requires comparisons with other possible models, Chapter 7 briefly describes several alternative approaches to the study of complex systems: cellular automata (cf. Wolfram 1984), classical linear stochastic models, and nonlinear time series analysis. Following these core chapters on the statistical foundations of fractal models, the next two chapters illustrate fractal modelling through both new and classical examples, presented at the level of journal articles. Chapter 8 begins with a new study of earthquake time series aimed at the possibility of predicting large events. Chapter 8 also summarizes two applications of fractals to developmental biology with a surprising common thread: the same mathematics governs the formation of patterns of islets of Langerhans and pancreatic ductules (recent results of Hastings *et al.* 1992) and of neuronal processes (Caserta *et al.* 1990; Kleinfeld *et al.* 1990). Chapter 9 concludes with suggested applications of Hurst's techniques to a wide range of weather phenomena, ranging from rainfall patterns to ocean surface temperatures and El Niño periodicity.

The culminating case studies of Part IV include 'behind the scenes' material cut from most research presentations: deliberations about the choice of models, historical and logical motivations, and, in particular, thorough discussions of paths not followed. Chapter 10 describes the fractal modelling of vegetative ecosystems (cf. Levin and Paine 1974; Hastings *et al.* 1982). Chapter 10 concludes with a list of open questions which might be at least partially answered with fractals: For example, are there general rules relating the number of species of a given size and the number of species found in a given area? Chapter 11 is our new attempt to forecast local extinction of small bird populations (cf. Sugihara and May, 1990*a*). This chapter concludes with a challenge to combine several partial measures for the persistence of small populations to form a practical tool for forecasting.

Finally, Chapter 12 of Part V is a toolbox of our programs, ready to use for both new and experienced users of fractals. The programs in Chapter 12 are written in Turbo Pascal, but no special features of Pascal are used. The reader with a knowledge of Basic can easily understand these programs and translate them if necessary. On the other hand, the reader with strong programming skills, including a knowledge of C, can easily combine the algorithms underlying these programs with programs in Press *et al.*'s (1989) *Numerical recipes* to obtain fast efficient code.

Legal issues. The authors and Oxford University Press explicitly disclaim any and all warranties, express or implied, concerning the merchantability, suitability or fitness of the software described herein. The user assumes all responsibility and risk associated with using such software. In particular, neither the authors nor Oxford University Press shall be responsible for damages of any kind, including but not limited to special, incidental or consequential damages, resulting from the use of the software.

This book can be read straight through as a text beginning with the examples of Chapter 1 and the basic mathematics of Chapter 2, or the reader can jump in at many points: the computational techniques of Chapters 3–5, the 'how to' statistics of Chapters 6 and 7, representative applications including new applications to the geometry of pancreatic islets, earthquake time series, and ocean temperature data in Chapters 8 and 9, or either of the case studies. The list of references includes eclectic comments and a few recommendations for further reading.

We have found the science of fractals both interesting and fun, and hope that the reader will agree.

1

Our view of nature

Natural patterns, especially those in ecosystems, frequently appear irregular, complex, and hard to measure, even at very small scales. Consider, for example, the problem of measuring the length of the coastline of England (Richardson 1961; cf. Mandelbrot 1977, 1982). Richardson attacked this question by traversing the coastline in a sequence of small steps, and counting the number of steps required. If the coastline could be traversed in $n(\Delta s)$ steps, each of length Δs, then the product $n(\Delta s)\Delta s$ would be a close approximation of the actual length. If the coastline were a smooth curve, then for very small steps the apparent length $n(\Delta s)\Delta s$ would closely approximate the actual length, implying that

$$n(\Delta s) = \text{const} \times (1/\Delta s).$$

However, Richardson found that the apparent length $(n(\Delta s)\Delta s$ appeared to increase without bound as Δs was decreased, and in fact that

$$n(\Delta s) = \text{const} \times (1/\Delta s)^{-D}, \tag{1.1}$$

where the exponent D was strictly greater than 1.

Similar regularities had been found in other attempts to measure complex patterns such as the distribution of areas of Aegean islands (Korcak 1938) and fluctuations in discharges from the Nile river (Hurst 1951, 1956; cf. Mandelbrot (1977, 1982) for both Korcak and Hurst). Korcak found that the distribution of sizes of Aegean islands followed the formula

$$n(a) = \text{const} \times a^{-B}, \tag{1.2}$$

where $n(a)$ denotes the number of islands of area at least a. In designing the Aswan dam, Hurst found that the difference between observed and expected cumulative discharges over time periods of duration Δt followed the formula

$$y = \text{const} \times \Delta t^{H}. \tag{1.3}$$

Formulae of the form

$$y = ax^{b} \tag{1.4}$$

are called *power laws*. Why do power laws appear in the measurement of such irregular natural patterns and time series?

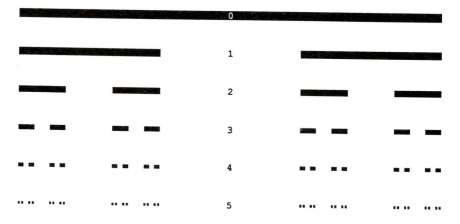

Fig. 1.1   Construction of the Cantor set. This complex pattern is built through the endless repetition of a very simple process, removing the middle third of a line segment to obtain two shorter segments.

Power laws are also found in Euclidean geometry, allometry, and statistics. For example, the familiar formulae for the area of a square and volume of a cube are the power laws

$$A = s^2 \quad \text{and} \quad V = s^3, \tag{1.5}$$

in which the scaling exponent is the dimension of the object. We shall see below that many complex patterns can be similarly characterized by an appropriate dimension. These formulae have simple explanations in cases where the parameter s is an integer. A square of side s can be decomposed into s^2 similar smaller unit squares, and a cube of side s can be decomposed into s^3 similar smaller unit cubes.

Such simple geometric principles underlie some of the power laws in allometry or the measurement of form in animals (cf. Peters 1983; Schmidt-Nielsen 1984). For example, for a family of similar animals, the surface area is proportional to the square of the length and the mass is proportional to the cube of the length. Combining these formulae yields the power law

$$\text{mass} = \text{const} \times (\text{surface area})^{3/2}. \tag{1.6}$$

This power law provides sharp limits on the form and metabolic requirements of many families of animals. A power law also connects the metabolic rate and weight of animals (cf. Peters 1983; Schmidt-Nielsen 1984):

$$\text{metabolic rate} = \text{const} \times (\text{weight})^{3/4}. \tag{1.7}$$

Although the power law (1.6) is a simple consequence of geometric formulae, the power law (1.7) reflects more complex scaling of metabolic processes.

Even random walks display regularities in the form of power laws. The Central Limit Theorem of statistics implies that the change in position Δs resulting from a random walk over a time interval Δt is approximately

$$\Delta s \approx \text{const} \times (\Delta t)^{1/2}. \tag{1.8}$$

Mandelbrot (1977, 1982) discovered that a natural geometry, the geometry of *fractals*, could provide a unified framework and explanation for many of these power laws. It is hard to measure the coastline of England because the coastline of England never appears straight, even on very small scales. It is similarly hard to measure the path of Brownian motion, the continuous-time analogue of a random walk (Perrin 1906; cf. Mandelbrot 1977, 1982) because irregularities persist, even on very small scales. However, a highly magnified view of a portion of the coastline of England *does* resemble the coastline, and a highly magnified view of a portion of the graph of Brownian motion *does* resemble the graph itself. Mandelbrot's surprising observation was that the coastline of England could be 'built' out of pieces similar to the whole coastline, and a continous-time random walk could be built of similar shorter random walks, both in close analogy to building a square out of similar smaller squares. Mandelbrot called this property *self-similarity* (he actually and correctly found only a weaker scale-invariant property, self-affinity, for the graph of Brownian motion), and called self-similar objects *fractals*. The measurement of fractals is no harder than the measurement of the regular objects of Euclidean geometry if the right building blocks, the fractals themselves, are used in measurement—*self-similarity forces the complexity of the object into the building blocks and describes the inherent regularities through power laws.*

Fractals thus provide a simple description of many natural forms. For example, a standard Euclidean description of the black spleenwort fern illustrated in the Introduction might require thousands of data points and fitted parameters (Barnsley 1988). In constrast, Barnsley describes this fern using a fractal building block, four transformations having only six parameters, and the iterative process of recursion.

In an abstract setting, recursion generates such 'mathematical monsters' (Mandelbrot, 1977, 1982) as Cantor (1872) sets, Brownian motion (Perrin 1906; cf. Mandelbrot 1982, pp. 6ff.), the Koch snowflake (von Koch, 1906), and Weierstrass's continuous nowhere-differentiable function (cf. du Bois, 1875). The graphs of Brownian motion, the Koch snowflake, and the Weierstrass function *never* look like straight lines, even on arbitrarily small scales, and thus cannot be measured by using sufficiently small line segments. Cantor (1872) described uncountably infinite sets of measure zero; these Cantor sets also challenged the classical concepts of measurement. Cantor sets are complex on all scales, and in fact their small-scale complexity mimics their large-scale complexity. They can all be easily described with iterative processes and measured with fractals.

For example, the first stage in the iterative construction of the Cantor set is just a line segment. Each successive stage is constructed from the previous stage by removing the (open) middle third of all segments in the previous stage (see Fig. 1.1). This figure suggests that a Cantor set constructed from an interval of length 3 can be decomposed into 2 smaller 'unit' Cantor sets, each constructed from a unit interval of length 1, and more generally that a Cantor set constructed from an interval of length 3^k can be decomposed into 2^k smaller 'unit' Cantor sets. At least for s of the form 3^k, a Cantor set

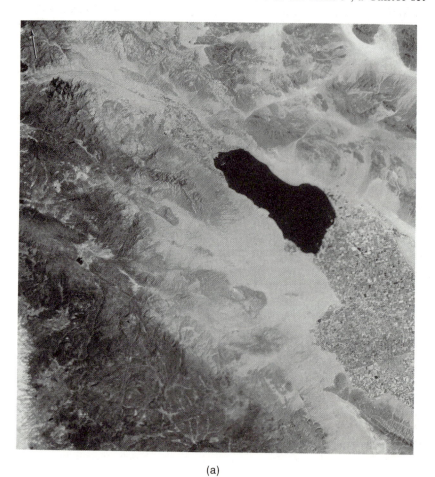

(a)

Fig. 1.2 A montage of natural and artificial patterns. (a) A false colour Landsat photograph of natural and agricultural patterns near Salton Sea in Southern California (cf. Short *et al.* 1976, plate 133, reprinted with permission of NASA). The natural patterns appear regular on sufficiently small scales. (b) Islands of vegetation (cypress and mixed cypress patterns) in the Okefenokee Swamp (Hastings *et al.* 1982, Fig. 3). (c) A computer generated landscape generated with

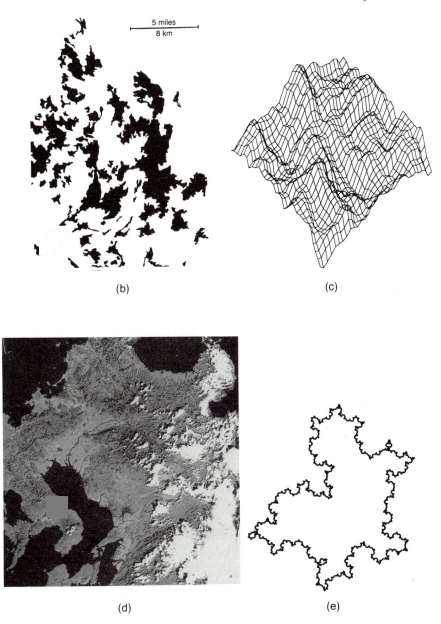

(b)

(c)

(d)

(e)

Fig. 1.2 (*continued*) Mandelbrot–Weierstrass fractals; see Sections 5.6 and 12.3. Islands like those in (b) can be generated by flooding the lower elevations of the landscape. (d) A Landsat photograph of northwestern Kyushu island in Japan (cf. Short *et al.* 1976, Plate 319, reprinted with permission of NASA). (e) A single computer-generated island, using the randomized Koch snowflake of Section 2.4 (see also Section 12.2).

constructed from an interval of length s can thus be measured in terms of unit Cantor sets, and has measure given by the power law

$$m = 2^k = s^{\log 2/\log 3}. \tag{1.9}$$

We shall see that many iterative growth, branching, and aggregation processes of biology give rise to power laws. Small-scale processes combine and form aggregated larger-scale processes in ecosystem dynamics. Iterated growth and branching rules can also give rise to complex patterns such as the vascular system and the bronchi of the lungs (cf. Mandelbrot 1977, 1982; Barnsley 1988). The abstract Brownian motion as well as actual cumulative river discharges are aggregations of smaller discharges. The dynamics behind the formation of islands, coastlines, and mountain ranges occurs on many scales. Species, community, and system dynamics in ecosystems represent successively higher aggregates of species-level dynamics.

Barnsley (1988) defines fractals in terms of iterating functions. We shall follow a more geometric approach because it is frequently difficult to find explicit algebraic representations of the iterative processes underlying natural dynamics, but shall make frequent reference to the use of power laws and fractals to distinguish between classes of underlying dynamics. Physicists call these classes 'universality' classes, and in fact the geometry of fractals owes much to the physical sciences.

The purpose of this book is to develop the science of random fractals from the viewpoint of applications to the natural sciences. We aim to be precise in the use of mathematics and statistics. Some proofs are outlined or included in order to make the ideas precise.

Part II begins with a chapter on the basic foundations of fractals: self-similarity, self-affinity, scaling and Hausdorff dimensions, and the mathematics of power laws, followed by three chapters of practical techniques for computing the dimension and other scaling exponents associated with fractal patterns and time series.

Part II

The mathematics of random fractals

The examples from Chapter 1 demonstrated that natural patterns can be extremely complex. However, these natural patterns appeared statistically scale-invariant, that is, statistically unchanged under magnification or contraction, at least over a fairly wide range of scales. Scale-invariant objects are called fractals; statistically scale-invariant objects are called random fractals. The next four chapters develop the mathematics of random fractals, with a special emphasis on measurement and description.

2

Fractals and power law scaling

2.1 Introduction

The examples from Chapter 1 demonstrated that some natural patterns can appear extremely complex. However, they may display an underlying simplicity through scale-invariance. Again, scale-invariance means that the pattern appears unchanged under magnification or contraction. More generally, there may be several scaling regions, separated by breakpoints, with scale-invariance holding within each region, but failing when a break-point is crossed.

In this chapter we shall define fractals as geometric objects which exhibit scale-invariance, and we shall show how scale-invariance leads to a class of scaling rules—power laws—characterized by scaling exponents. These scaling exponents are constant within each scaling region, but jump at the breakpoints separating scaling regions.

We consider a pattern or object to be scale-invariant within a scaling region if the pattern or object contains no natural internal measures of size, and thus appears the same at all scales within the scaling region. For example, a set of islands may be scale-invariant despite the fact that we can measure the area or perimeter of each island (see Fig. 1.2 above). In this case, scale-invariance simply means that the small islands are essentially reduced versions of the large islands, and the large islands enlarged versions of the small islands. We might then consider the relationship between the area and the perimeter of each island in the set. As explained formally in Section 2.6, scale-invariance is manifest algebraically in a power law rule for the distribution of areas of islands,

$$N(a) = \text{const} \times a^{-B}, \qquad \cdot \qquad (2.1)$$

where $N(a)$ denotes the number of islands of area at least a (Korcak 1938), another power law relating the area a and perimeter p of each island,

$$p = \text{const} \times a^{D/2}, \qquad (2.2)$$

and other power laws described below.

The first three sections develop the concepts of self-similarity, regular fractals, and random fractals. In particular, Section 2.4 contains an axiomatic characterization of an important class of prototype random fractals: the

graphs of random walks, their diffusion limits, and their fractal generalizations. In Section 2.5 we introduce the use of the scaling dimension and the
more general Hausdorff dimension to describe the scaling properties of
regular and random fractals. The parameter D in formula (2.2) is in fact the
Hausdorff dimension of the boundaries of the islands. Chapters 3–5 develop
a wide variety of methods for computing the Hausdorff dimension and
related scaling exponents for fractal patterns and functions. Finally, Section
2.6 introduces algebraic self-similarity and power laws.

2.2 Self-similarity and fractals

Fractals are defined to be scale-invariant (self-similar or self-affine) geometric
objects. A geometric object is called *self-similar* if it may be written as a
union of rescaled copies of itself, with the rescaling isotropic or uniform in
all directions. A geometric object is called *self-affine* if it may be written as
a union of rescaled copies of itself, where the rescaling may be anisotropic
or dependent on the direction. *Regular fractals* display exact self-similarity.
Random fractals display a weaker, statistical version of self-similarity or, more
generally, self-affinity. Although virtually all natural fractals are random, the
concept of self-similarity is best first explored through the study of regular
fractals.

The class of regular fractals includes many familiar simple objects such as
line intervals, solid squares, and solid cubes (because of the generality of our
definition), and also includes many irregular objects such as the Cantor set
and Koch snowflake. The scaling rules are characterized by scaling exponents
(dimension). We shall see that 'simple' regular fractals have integral scaling
dimensions, and complex self-similar objects such as the Cantor set and Koch
snowflake shown in Fig. 2.1 have nonintegral dimensions.

We shall develop a more precise definition of statistical self-similarity and
self-affinity as we develop the theory of random fractals. For now it suffices
to state that a geometric object is statistically self-similar or self-affine,
respectively, if it is the union of several pieces, each of which is statistically
an appropriately rescaled copy of itself. The randomized Koch snowflake
shown in Fig. 1.2 and discussed in Section 2.4.2 is a typical random fractal.

2.3 Regular fractals

In this section we shall develop the concepts of self-similarity and regular
fractals. Many regular geometric figures are self-similar and self-similarity
has its origins in the formulae for the area of a square, $A = s^2$, and volume
of a cube, $V = s^3$. If s is a whole number, then these formulae follow from
decomposing the square of side s into s^2 similar unit squares and from

(a)

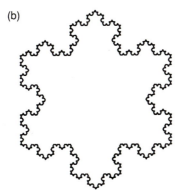

(b)

Fig. 2.1 (a) The Cantor set and (b) Koch snowflake: mathematical monsters of the nineteenth century tamed by fractal geometry.

similarly decomposing the cube of side s into s^3 similar unit cubes. (This type of decomposition can be made precise through the concept of an *almost-disjoint union*. The line segment of length (or side) s may be similarly decomposed into s segments each of unit length. Area and volume formulae thus relate the size of these simple objects to their linear scales, with the dimension serving as the key parameter. The formulae follow from decomposing the objects into unions of similar smaller objects—therefore the cube, square, and line segment are called *self-similar* and the dimension is called a *scaling* or *similarity* dimension. A point is trivially self-similar with scaling dimension 0.

2.3.1 *Construction of the Cantor set*

The power of the concepts of self-similarity and scaling dimension becomes apparent when they are used to characterize such seemingly complex and irregular objects as the Cantor set. Cantor (1872) constructed this set as a pathological example to demonstrate the need for careful hypotheses in mathematical analysis. The Cantor set is constructed as a limit of an iterative process in which the first stage, stage number 0, is the closed unit interval. Referring to Fig. 1.1 above, given any stage n, the next stage $n + 1$ is constructed by deleting the open middle third of each closed interval of stage n. The Cantor set is the resulting limit, and may be formally defined as the intersection of all of the approximating stages. Note that stage n is the union of 2^n closed intervals, each of length $1/3^n$. Thus stage n has length or linear measure $(2/3)^n$, which approaches 0 as n approaches infinity.

Although it is difficult to conceive of the Cantor set as a union of familiar objects from Euclidean geometry, it can be written as the union of smaller Cantor sets just as a square can be written as a union of smaller squares. In fact, by construction, the Cantor set is the union of 2 smaller Cantor sets, each obtained by contracting the original Cantor set by a factor of 3. More generally, for any number m which is a power of 2,

$$m = 2^n, \tag{2.3}$$

the Cantor set is the union of m smaller Cantor sets, each a factor of 3^n smaller than the original Cantor set. Since a Cantor set of scale 3 is a union of 2 Cantor sets of scale 1, and

$$2 = 3^{\log 2/\log 3}, \tag{2.4}$$

we assign the Cantor set a scaling dimension of $\log 2/\log 3 = 0.63 \dots$. Unlike the case of the square and similar 'simple' geometric figures discussed above, the scaling dimension of the Cantor set is not an integer.

There is a second standard description of the Cantor set which sheds additional light on its scaling dimension. This description involves *ternary decimals*, where a ternary decimal is the analogue of a decimal in base three notation. That is, a ternary decimal is a sequence of the form

$$a_1 a_2 a_3 \dots \tag{2.5}$$

which represents the number

$$a_1/3 + a_2/3^2 + a_3/3^3 + \cdots . \tag{2.6}$$

By analogy with the usage of ordinary decimals, ternary decimals are generally written in the form

$$0.a_1 a_2 a_3 \dots . \tag{2.7}$$

We note that the equivalence

$$0.999 \dots = 1 \tag{2.8}$$

for ordinary decimals corresponds to the equivalence

$$0.222 \dots = 1 \tag{2.9}$$

for ternary decimals, and adopt the convention that any *finite* ternary decimal

$$0.a_1 a_2 a_3 \dots a_n, \quad a_n \neq 0, \tag{2.10a}$$

will always be written as an eventually *repeating* ternary decimal

$$0.a_1 a_2 a_3 \dots a_n' 222 \dots , \quad a_n' = a_n - 1. \tag{2.10b}$$

Suppose that all finite ternary decimals are represented as such eventually repeating ternary decimals. Then stage 0 of the Cantor set consists of all ternary decimals, stage 1 of the Cantor set consists of all ternary decimals with $a_1 \neq 1$, and stage 2 of the Cantor set consists of all ternary decimals with $a_1 \neq 1$ and $a_2 \neq 1$. More generally, each 'middle third' corresponds to a '1' in the appropriate position in a ternary decimal expansion. Therefore the Cantor set consists of all ternary decimals which can be written with no 1's. Reducing the scale of the Cantor set by $1/3^n$ corresponds to multiplying the ternary decimal expansions by

$$1/3^n = 0.000 \dots 01 \qquad (n-1 \text{ zeros}), \qquad (2.11a)$$

and thus the mapping

$$0.a_1a_2a_3 \dots \rightarrow 0.000 \dots 00a_1a_2a_3 \dots \qquad (n \text{ zeros}). \qquad (2.11b)$$

Moreover, one may obtain the 2^n copies of the reduced Cantor set which make up the original Cantor set by replacing none, any, or all of the first n zeros by 2's. It is easy to see that this replacement may be done in 2^n ways. This shows rigorously that the original Cantor set is the union of 2^n such reduced Cantor sets.

Moreover, this description provides a simple way to generate a Poisson distribution on the Cantor set. Define a ternary decimal $a = 0.a_1a_2a_3 \dots$ as follows. For each i, let

$$a_i = \begin{cases} 0 & \text{with probability } \tfrac{1}{2}, \\ 2 & \text{with probability } \tfrac{1}{2}. \end{cases} \qquad (2.12)$$

It is easy to see that

$$Pr(0.a_1a_2a_3 \dots a_n0 \leqslant 0.a_1a_2a_3 \dots a_na_{n+1} \dots \leqslant 0.a_1a_2a_3 \dots a_n2) = 1/2^n, \qquad (2.13)$$

and that this is the same as the probability that the point $0.a_1a_2a_3 \dots a_na_{n+1} \dots$ lies in the corresponding reduced Cantor set.

We now describe a second regular fractal, the Koch snowflake, and relate its iterative construction to its scaling dimension, just as we did for the Cantor set. The construction starts with a triangle as stage 0. Stage $n+1$ is constructed from stage n by replacing each line segment of stage n by a polygonal path consisting of four line segments, each one-third the length of the original segment (see Fig. 2.2).

Although the Koch snowflake is not itself the union of similar smaller snowflakes, by construction, each side of the Koch snowflake is such a union of 4 similar smaller curves, each one-third of the length of the original side. Moreover, by construction, each reduced side has the same scale-invariant property, and thus the scaling dimension of the Kock snowflake is log 4/log 3, or about 1.26.

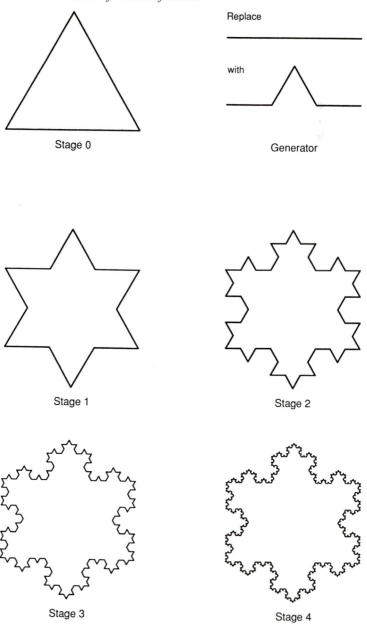

Fig. 2.2 Construction of the Koch snowflake. The Koch snowflake is constructed as the limit of a sequence of simple iterative steps. Starting with the equilateral triangle at the top left, each successive stage is constructed by replacing line segments with copies of the polygonal generator at the top right. Compare Fig. 1.1 (construction of the Cantor set) and Fig. 2.5 (construction of the randomized Koch snowflake).

Natural branching processes such as the formation of neuronal processes, the bronchial tubes, the vascular system, and pancreatic ducts reflect a random version of the above iteration (see Chapter 8). Mandelbrot–Weierstrass fractals (Berry and Lewis 1980) display a wave upon wave appearance reminiscent of the appearance of successive stages in the construction of the Koch snowflake. Chapter 12 contains programs for construction of the Koch snowflake and Mandelbrot–Weierstrass and related fractals.

REMARKS 2.1 The Cantor set, which consists of a totally disconnected set of points (cf. Hurewicz and Walman 1941, Dugundji 1966 or any text on point set topology for definitions), has topological dimension 0, strictly less than its scaling dimension $\log 2/\log 3 = 0.63 \ldots$ (see Figs. 2.6 and 3.2 below). Similarly, the Koch snowflake is a Jordan curve in the plane and has topological dimension 1, also strictly less than its scaling dimension $\log 4/\log 3 = 1.26 \ldots$. In fact, unlike us, Mandelbrot (1977, 1982) includes the condition that the scaling dimension exceed the topological dimension in his definition of a fractal.

We shall return to these ideas in Section 2.4, and conclude this section with a precise definition of almost-disjoint union.

DEFINITION 2.1 A set X is the *almost-disjoint* union of two sets A and B if X is the union of A and B, and the intersection of A and B has lower dimension than the dimensions of A and of B. Almost-disjoint unions of more than two sets are defined similarly. Figure 2.3 illustrates this concept.

DEFINITION 2.2 Suppose that a self-similar set X is the almost-disjoint union of n copies of X, each contracted by a factor k and translated by a

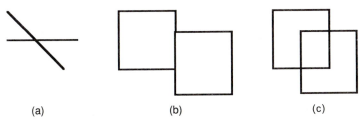

(a) (b) (c)

Fig. 2.3 Almost-disjoint union. (a) An almost-disjoint union of two lines—the lines (dimension 1) meet in a point (dimension 0). (b) An almost-disjoint union of two squares—the squares (dimension 1) meet in a line (dimension 1). (c) A union of two squares which is clearly not an almost-disjoint union—the squares (dimension 1) meet in a rectangle (dimension 1).

vector a_i in space:

$$X = \bigcup_{1 \leqslant i \leqslant n} (1/k)X + a_i. \tag{2.14}$$

Then X has scaling dimension $\log n / \log k$.

2.4 Random fractals

The graph of Brownian motion, a continuous-time random walk (shown in Fig. 2.4 below), is the prototype random fractal. We therefore begin with the construction of random walks, and their diffusion limits, Brownian motion. This construction can be found in Lin and Segel (1974), Hogg and Tanis (1977), or any elementary textbook on probability and statistics. We shall then describe fractal generalizations of Brownian motion, and conclude with their axiomatic characterization (Mandelbrot, 1977, 1982).

A random walk in one dimension is defined as follows. Choose a time step Δt and a space step Δy. Let

$$y(0) = 0, \tag{2.15}$$

and define $y(t)$ inductively for times t which are a whole-number multiple of the time step Δt, by the formula

$$y(t + \Delta t) = y(t) + \Delta y(t), \tag{2.16}$$

where

$$\Delta y(t) = \begin{cases} \Delta y & \text{with probability } \tfrac{1}{2}, \\ -\Delta y & \text{with probability } \tfrac{1}{2}, \end{cases} \tag{2.17}$$

independently of any previous steps. Thus each random walk is a function defined at the points $\{n\Delta t : 0, 1, 2, ...\}$ which takes values in the set $\{n\Delta y : n = 0, \pm 1, \pm 2, ...\}$. Each random walk may be considered as a sample from the space of all random walks. The statistics of random walks may be readily computed, since the value of a random walk at time $n\Delta t$ is just the sum of n independent identically distributed random variables ΔY_i, each with distribution

$$\begin{cases} \Delta y & \text{with probability } \tfrac{1}{2}, \\ -\Delta y & \text{with probability } \tfrac{1}{2}. \end{cases} \tag{2.18}$$

We now calculate the first and second moments of $y(t)$. Recall the following properties of the expectation function E. First, the expectation is linear, which means that, for any constant c and any two random variables X_1 and X_2,

$$\mathrm{E}(X_1 + X_2) = \mathrm{E}(X_1) + \mathrm{E}(X_2) \quad \text{and} \quad \mathrm{E}(cX_1) = c\mathrm{E}(X_1). \tag{2.19}$$

In addition, if the random variables X_1 and X_2 are *independent*, then

$$E(X_1 X_2) = E(X_1)E(X_2). \tag{2.20}$$

We shall informally write $E(x)$ where x is a random sample from a distribution in place of the more formal use of $E(X)$.

PROPOSITION 2.1 For any time t, the expected value (or first moment) of $y(t)$ is 0.

Proof. By definition, $E(y(0)) = E(0) = 0$, and $E(\Delta y(t)) = 0$ for all times t. Now assume inductively that

$$E(y(t)) = 0 \quad \text{for } t = n\Delta t.$$

Then because the expected value of a sum is the sum of the respective expected values, the inductive construction of a random walk implies that

$$E(y(t + \Delta t)) = E(y(t) + \Delta y(t)) = E(y(t)) + E(\Delta y(t)) = 0.$$

Therefore,

$$E(y(t)) = 0 \quad \text{for } t = (n + 1)\Delta t.$$

The conclusion follows by mathematical induction.   □

PROPOSITION 2.2 For any time t, the expected value of $[y(t)]^2$ (the second moment or, since $E(y(t)) = 0$, the variance of $y(t)$) is $n\Delta t^2$.

Proof. We compute similarly, beginning with the calculations that $E([y(0)]^2) = E(0) = 0$, and $E([\Delta y(t)]^2) = E(\Delta t^2)$ for all times t. As above, assume inductively that

$$E([y(t)]^2) = n\Delta t^2 \quad \text{for } t = n\Delta t.$$

Then, since all increments are independent, $y(t)$ and $\Delta y(t)$ are also independent by construction, which implies that

$$E([y(t + \Delta t)]^2) = E([y(t)]^2) + 2E(y(t))E(\Delta y(t)) + E([\Delta y(t)]^2)$$

$$= E([y(t)]^2) + \Delta y^2. \tag{2.21}$$

By equation (2.21),

$$E([y(t)]^2) = n\Delta t^2 \quad \text{for } t = (n + 1)\Delta t.$$

The conclusion follows by mathematical induction.   □

Thus the expected value of the *square* of the distance traversed by a random walk grows linearly with the time. A rough translation is that the displacement grows as the square root of the time. This translation is readily made precise by using the standard deviation, or root mean square displacement, as a kind of 'average'.

2.4.1 *The asymptotic distribution of* y

The central limit theorem of statistics implies that the sum of n independent identically distributed bounded random variables approaches a normal (or Gaussian) distribution. Therefore, for large n, the distribution of $y(n\Delta t)$ is asymptotically normal with mean 0 and variance distribution $n\Delta y^2$, denoted $N(0, n\Delta y^2)$.

In order to introduce the continuous version of random walks, Brownian motion, we further develop their long-term scaling behaviour. Consider two random walks, with respective time and space steps Δt_i and Δy_i. Suppose that the time steps are commensurable and that, at some time t,

$$t = n_1\Delta t_1 = n_2\Delta t_2. \tag{2.22}$$

Then, at this time t, the position of the first random walk is approximately (asymptotically as $n \to \infty$) a sample from the normal distribution $N(0, n_1\Delta y_1^2)$ and the position of the second random walk is approximately a sample from the normal distribution $N(0, n_2\Delta y_2^2)$. However, if

$$n_1\Delta y_1^2 = n_2\Delta y_2^2, \tag{2.23}$$

then the two normal distributions are the same:

$$N(0, n_1\Delta y_1^2) = N(0, n_2\Delta y_2^2). \tag{2.24}$$

Equations (2.23) and (2.24) imply that, for times long compared with the time step Δt, the statistics of a random walk depends only on the ratio

$$\Delta y^2/\Delta t. \tag{2.25}$$

Thus, we may use any time step Δt much less than any time scale under consideration in formulating and applying random walk models. In a formal sense, we take the *diffusion limit* of random walks by letting the time step Δt approach 0, and requiring that the ratio $\Delta y^2/\Delta t$ of equation (2.25) approach a constant called the diffusion rate R. Letting the time step Δt approach 0 makes any constant time scale long, and thus yields a continuous-time version of random walks called Brownian motion (see Lin and Segel 1974, for details). The effect of diffusion limits is illustrated in the random walk program (Section 12.4.1).

REMARKS 2.2 We shall use the scaling rule for Brownian motion, formula (2.25) above, in Chapter 3 to show that its graph has scaling dimension 1.5.

We conclude this section by giving the axiomatic characterization of

Brownian motion and its fractal generalization following Mandelbrot (1977, 1982).

DEFINITION 2.3 A continuous process $\{y(t)\}$ is called a *continuous-time random walk* or a *Brownian process* if, for any time step Δt, the increments $\Delta y(t) = y(t + \Delta t) - y(t)$ are

(i) Gaussian,
(ii) of mean 0, and
(iii) variance proportional to Δt.

In the presence of axiom (ii), axiom (iii) is equivalent to axiom

(iv) successive increments $\Delta y(t)$ and $\Delta y(t + \Delta t)$ are uncorrelated.

The axioms which characterize random walks can be readily generalized to characterize *fractal processes* (Mandelbrot, 1977, 1982) by introducing an additional parameter, the 'Hurst exponent' H $(0 < H < 1)$ and replacing axom (iii) by the axiom

(iii') variance proportional to Δt^{2H}

(Hurst 1951, 1956 first observed similar scaling properties; cf. Mandelbrot 1977, 1982). A random walk has Hurst exponent $H = \frac{1}{2}$. As above, axiom (iii') is equivalent to a simple axiom (iv') about the correlation of successive increments: in a fractal process successive increments are correlated with coefficient of correlation ρ, independent of the time step h, where ρ is defined by the formula

$$2^{2H} = 2 + 2\rho \quad (-\tfrac{1}{2} < \rho < 1). \tag{2.26}$$

The axioms characterize the scaling behaviour of fractal processes. If $\{y(t)\}$ is a fractal process with Hurst exponent H, then, for any constant $c > 0$, the process

$$y_c = (1/c^H)y(ct) \tag{2.27}$$

is another fractal process with the same statistics. Physicists call this rescaling *renormalization*.

In order to introduce the concept of random fractals, consider the family F_H of graphs of all fractal processes of Hurst exponent H. The family F_H is closed under the renormalization (2.27) and all elements of F_H share the same statistical properties. More generally, a *random fractal* is an element of a set S which is closed under application of a renormalization formula, or a group of renormalization formulae. This is essentially Barnsley *et al.*'s (1986) and Barnsley's (1988) definition of fractals through iterated function systems. There are 'trivial' examples, such as the set of all curves in the plane, but most interesting examples, such as the graphs of fractal processes, are characterized by a set of statistical properties.

For example, one can define fractal functions of two or more variables similarly to fractal processes. For a fractal function $z = f(x, y)$ of two variables the increments $\Delta_x z = z(x + \Delta x, y) - z(x, y)$ and $\Delta_y z = z(x, y + \Delta y) - z(x, y)$ satisfy axioms (i)–(iv). In fact, increments in any direction satisfy the same axioms, and the intersection of a *fractal sheet* (the graph of a fractal function of two variables) with a vertical plane is a fractal curve with the same exponent. It is easy to see that fractal sheets are also random fractals.

We conclude by asking what is the dimension of the graph of Brownian motion $y = B(t)$? Can it be computed from the renormalization property? In order to have a bounded set, we restrict the domain of t to the closed unit interval $[0, 1]$. The graph of $y = B(t)$ $(0 \leqslant t \leqslant 1)$ is the almost-disjoint union of four segments, each obtained by a further restriction of the domain to one-quarter of the unit interval. However, by renormalization, the original graph is not self-similar but rather only self-affine: statistically each segment is reduced by a factor of $\frac{1}{4}$ horizontally but only $\frac{1}{2}$ (*the square root of* $\frac{1}{4}$) vertically (see Fig. 2.4). Moreover, if contracting both axes by a factor of 4 led to a similar graph, then D would be 1, despite the apparent non-Euclidean complexity of the graph. However, we shall see in Section 3.2 below, using an appropriate variant of the Hausdorff dimension, that the missing factor of 2 in the contraction along the y-axis adds an additional $\log 2/\log 4$ to the dimension, making $D = 1.5$.

2.4.2 A randomized Koch snowflake

The relationship between regular and random fractals can be understood by considering another example, a 'randomized' Koch snowflake in which the bumps move up or down. More formally, a randomized Koch snowflake has two generators, shown in Fig. 2.5 below. Stage $n + 1$ is constructed from stage n by replacing each line of stage n by one of the two generators, chosen randomly and independently with equal probability, from the preceding iteration. Figure 2.5 also illustrates the construction of a randomized Koch snowflake. (The self-intersections can be avoided by slightly shrinking the bumps in the generators.) It is easy to see that each side of the randomized Koch snowflake is the union of 4 smaller statistically similar curves, each contracted by a factor of $\frac{1}{3}$ from the original side. Thus, as in Fig. 2.2 and the subsequent discussion, the randomized Koch snowflake has the same fractal dimension, $\log 4/\log 3$, as the corresponding regular Koch snowflake.

We shall next discuss the definition of dimension and the algebra of power laws, or scaling rules, parametrized by scaling exponents.

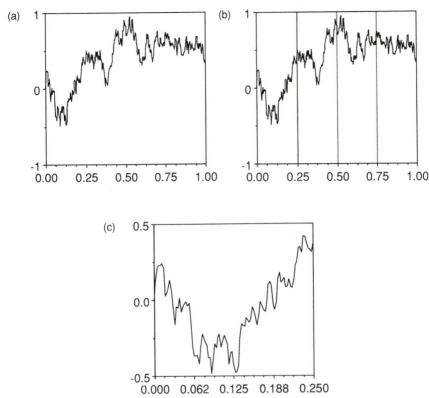

Fig. 2.4 Scaling in Brownian motion. (a) Graph of a typical Brownian function $y = B(t)$ on the unit interval $0 \leqslant t \leqslant 1$. (b) The graph consists of four subgraphs—each obtained by restricting the domain of the original graph to one of the subintervals $0 \leqslant t \leqslant \frac{1}{4}, \frac{1}{4} \leqslant t \leqslant \frac{1}{2}, \frac{1}{2} \leqslant t \leqslant \frac{3}{4}, \frac{3}{4} \leqslant t \leqslant 1$. (c) Graph of a rescaled version of the first subgraph in (b)—stretching the t-axis by a factor of 4 and the y-axis by a factor of 2 yields the Brownian function $y = \frac{1}{2}B(4t)$ whose graph is statistically similar to the graph in (a).

2.5 Dimension

In this section we extend and formalize the concept of scaling dimension through the concept of *Hausdorff dimension*, and compare and contrast the Hausdorff dimension with the usual topological dimension. (The properties of dimension functions will be given in Section 3.7.) We shall see that the Hausdorff dimension is a natural generalization of the scaling dimension defined above. There is one key difference. The scaling dimension measures the 'mass' (natural measure) of a self-similar set X in terms of small-scale copies of X. The Hausdorff dimension measures the mass of X in terms of Euclidean building blocks: open balls of a given radius. Although the

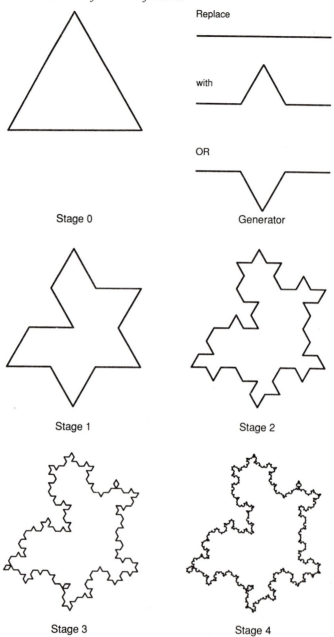

Replace

with

OR

Stage 0

Generator

Stage 1

Stage 2

Stage 3

Stage 4

Fig. 2.5   Construction of the randomized Koch snowflake. Like the original Koch snowflake of Fig. 2.2, the randomized version is constructed as the limit of a sequence of simple iterative steps. Starting with the equilateral triangle at the top left, each successive stage is constructed by replacing line segments with copies of one of the two polygonal generators, chosen at random, shown at the top right.

generality of the Hausdorff dimension requires a more complex definition, this generality itself leads to many practical computationally equivalent techniques (see Chapter 3).

The Hausdorff and 'usual' topological dimensions are both defined in terms of coverings by open sets and agree for nice Euclidean sets such as manifolds (cf. Hurewicz and Wallman 1941; Mandelbrot 1977, 1982). However, in general the Hausdorff dimension need not agree with the topological dimension (see also Remarks 2.1 above). We begin by recalling the definition of open set of Euclidean space.

DEFINITION 2.4 The *open ball* B(p, r) of radius r about the point p in Euclidean space is the set

$$B(p, r) = \{x : \operatorname{dist}(x, p) < r\},$$

where dist(x, p) is the distance between the points x and p. A set U in Euclidean space is called an *open set* if U is the union of a distance $r > 0$ such that the open ball B(p, r) is contained in U. A family of open sets $\{U_a\}$ is called an *open cover* of a set X if X is contained in the union $\bigcup_{\{a\}} U_a$ of the sets U_a.

We shall usually consider open coverings by open balls or open boxes (boxes without their boundaries).

Both the Hausdorff dimension and the topological dimension are defined by the properties of suitably minimal open coverings.

DEFINITION 2.5 The *topological dimension* of an object X in Euclidean space is defined as follows (cf. Hurewicz and Wallman 1941, Dugundji 1966, or any text on elementary point set topology). Consider a family of open sets (such as open boxes) which covers the object X in the sense that X is contained in the union of these open sets. A *refinement* of such an *open cover* is a second open cover each of whose open sets is contained in an open set of the given open cover. The topological dimension of an object is defined to be D_{top} provided that any open covering of the object admits a refinement in which any intersection of more than $D_{\text{top}} + 1$ distinct open sets is empty (see Fig. 2.6).

DEFINITION 2.6 The Hausdorff dimension D of a subset X of Euclidean space arises from asking 'how big is X?' for very general sets. The answer comes from counting the number of open balls needed to cover the set X. For each $r > 0$, let $N(r)$ denote the smallest number of open balls of radius r needed to cover X. One can show that the limit

$$D = \lim_{r \to 0} \left(-\log N(r) / \log r \right) \tag{2.28}$$

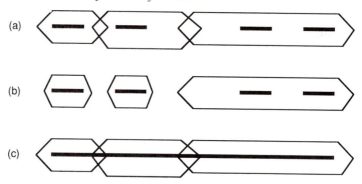

Fig. 2.6 Topological dimension of the Cantor set and the line. (a) A typical covering of the Cantor set by open sets. (b) Refinement of the covering in (a) so that no two open sets in the refinement intersect. Thus the Cantor set has topological dimension 0. (c) A typical covering of a line segment by open sets. Any such covering can be refined so that no three open sets intersect, but some pairs of open sets will always intersect unless the covering consists of only one open set. Thus a line segment has topological dimension 0. Compare Fig. 3.2 on the box dimension.

exists. The value of D is called the Hausdorff dimension of X. (Since $\log r \to -\infty$, the negative sign is needed in order that D be positive.)

REMARKS 2.3

(a) Formula (2.28) is equivalent to the approximate power law

$$N(r) \approx \text{const} \times r^{-D}. \tag{2.29}$$

We shall read formula (2.29) as '$N(r)$ scales asymptotically as r^{-D}' or loosely as '$N(r)$ scales as r^{-D}'. Roughly, two quantities x and y are *asymptotic as x approaches* 0 if the limit

$$\lim_{x \to 0} \log y / \log x \tag{2.30}$$

exists (cf. Lin and Segel 1974). We shall not need the precise definition. The concept of *asymptotic as x approaches* ∞ is defined analogously.

(b) The Hausdorff dimension is due to Carathéodory (1914) and Hausdorff (1919). Mandelbrot (1977, 1982) provides a nice survey of the Hausdorff dimension, its properties, and a list of references. The list includes Hurewicz and Wallman (1941), Billingsley (1967), Rogers (1970), and Adler (1981).

(c) The generality of the Hausdorff dimension makes it difficult to compute and to determine its properties (cf. Mandelbrot 1977, 1982). We therefore develop practical alternatives in Chapter 3.

PROPOSITION 2.3 Let X be a subset of Euclidean space with scaling dimension D. Then X also has Hausdorff dimension D.

Sketch of proof. First, assume that X can be decomposed into n rescaled copies of itself, each contracted by a linear factor of k, and thus that

$$D = \log n/\log k. \tag{2.31}$$

This assumption will be relaxed below. Choose a small $r_0 > 0$ and suppose that X can be covered by $N(r_0)$ open balls of radius r_0. Each reduced copy can clearly be covered by $N(r_0)$ 'rescaled' open balls of radius r_0/k. The union of the n families of open balls used to cover the n rescaled copies of X covers X, and thus X can be covered by $nN(r_0)$ open balls of radius r_0/k. Thus, at least approximately,

$$N(r_0/k) = nN(r_0). \tag{2.32}$$

The above construction can be iterated, obtaining the formulae

$$N(r_0/k^m) = n^m N(r_0) \quad (m = 1, 2, 3, ...). \tag{2.33}$$

Formulae (2.33) imply that

$$\lim_{m \to \infty} [-\log N(r_0/k^m)/\log(r_0/k^m)]$$

$$= \lim_{m \to \infty} [-\log n^m N(r_0)/\log(r_0/k^m)]$$

$$= \lim_{m \to \infty} [-(\log n^m + \log N(r_0))/(\log r_0 - \log k^m)]$$

$$= \lim_{m \to \infty} [(m \log n + \log N(r_0))/(m \log k - \log r_0)]$$

$$= \lim_{m \to \infty} [(\log n + (1/m) \log N(r_0))/(\log k - (1/m) \log r_0)]$$

$$= \log n/\log k$$

$$= D, \tag{2.34}$$

as required. Three technical problems must be handled in order to turn formula (2.34) into a mathematically rigorous proof. We shall outline their solution in the Appendix at the end of this Chapter. $\square$

REMARKS 2.4 Although the point, line, square, and cube have integral scaling dimensions, the irregular Cantor set and Koch snowflake, with dimensions $\log 2/\log 3$ and $\log 4/\log 3$, respectively, do not. The nonintegral dimensions both locate the figures in size and describe the irregularities. For

example, in terms of Hausdorff dimension, the Cantor set is intermediate
between a finite set of points (dimension 0) and a line segment (dimension 1).

In contrast the topological dimension of the Cantor set is 0 since the
Cantor set may be covered by a disjoint family of arbitrarily small open
intervals. The box dimension (see Fig. 3.2) and related dimensions differ
from the topological dimension in that the topological dimension does not
involve the concept of length or scaling. The box dimension is always at
least as large as the topological dimension whenever the box dimension is
defined. Since we are interested in scaling properties, the topological
dimension is not adequate for our purposes. Similarly the Koch snowflake
is intermediate between a regular polygon (dimension 1) and a filled-in
planar figure (dimension 2). The highly irregular space filling curves in the
plane have dimension 2.

2.6 Power laws

In this section we show that relationships between measurements of scale-
invariant systems take the form of power laws and give several examples.
For example, consider a pattern of separate islands, in which the ith island
has area x_i and perimeter y_i. Suppose the pattern is enlarged slightly. If the
pattern is scale-invariant, the area of each island will be multiplied by a
factor a, independent of the size of the island, and similarly the perimeters
will each be multiplied by a factor b, independent of the size of the island.
Moreover, by scale-invariance, the new (enlarged) islands will be similar to
the old (not enlarged) islands, so that any relationship between areas and
perimeters of old islands will still hold after enlargement. Since the pattern
is scale-invariant, we may repeat this process, with ax_0 replaced by $a^2 x_0$ and
by_0 replaced by $b^2 y_0$. Continuing inductively, suitably scaled versions of the
pattern have measurements

$$x = a^k x_0 \quad \text{and} \quad y = b^k y_0 \qquad (k = 0, 1, 2, ...). \qquad (2.35)$$

This inductive process is reminiscent of the scaling of iterative processes
in nature such as the branching of small blood vessels from larger ones (cf.
Mandelbrot 1977, 1982), and more generally of the iterated function systems
of Barnsley *et al.* (1986) and Barnsley (1988).

Formula (2.35) can also be shown to hold for negative exponents k.
Moreover, one can then obtain formula (2.35) first for all rational exponents
k and then for all exponents k by continuity. It follows from (2.35) that

$$\log y = k \log b + \log y_0 \quad \text{and} \quad \log x = k \log a + \log x_0. \qquad (2.36)$$

Formula (2.36) readily implies that

$$k = \log x / \log a - \log x_0 / \log a, \qquad (2.37)$$

and substituting (2.37) into the first formula in (2.36) yields

$$\log y = (\log b/\log a) \log x + [\log y_0 - (\log b/\log a) \log x_0]. \quad (2.38)$$

If we set $c = \log b/\log a$, it follows that

$$y = (y_0/x_0^c)x^c, \quad (2.39)$$

and thus the scale-invariant behaviour yields the scaling rule (2.39) for areas and perimeters, parametrized by the area–perimeter exponent

$$c = \log b/\log a \quad (2.40)$$

Any two measurements from a scale-invariant pattern show a similar exponential relationship. For example, the cumulative frequency exponent $-B$ parametrizes the relationship between an area a, and the number of islands $n(a)$ of area at least a:

$$n(a) = \text{const} \times a^{-B} \quad (2.41)$$

(cf. Korcak 1938). The negative sign is used so that the parameter B will be positive. Distributions of the form (2.41) are called *hyperbolic* by analogy with the rectangular hyperbola $y = 1/x$.

PROPOSITION 2.4 Functions f whose graphs appear the same on all scales must take the form $y = f(x) = \text{const} \times x^c$ for some exponent c.

Sketch of proof. First note that scale-invariance requires that

$$f(ax) = bf(x) \quad (2.42)$$

for any constant a and a related constant b which depends upon a. As in the discussion of the area–perimeter exponent, scale-invariance implies that

$$f(x) = \text{const} \times x^c, \quad (2.43)$$

where $c = \log b/\log a$, as required.   □

We shall call any exponent which measures a scaling behaviour of a function or geometric object, and is itself invariant with respect to that scaling behaviour, a *scaling exponent*. The scaling dimensions of Sections 2.2–2.4 are scaling exponents.

We conclude with an obvious but key result.

PROPOSITION 2.5 The log transformation of the power law (2.43) is the linear function

$$\log y = \log(\text{const}) + c \log x. \quad (2.44)$$

Log transforms play a central role in computing fractal exponents because of the central role of linear functions in mathematics and statistics. In particular, *linear regression* can be used to fit *log-transformed* power laws to *log-transformed* experimental data (cf. Korcak 1938, Chapter 6 and subsequent chapters). The next three chapters describe a wide variety of useful scaling exponents and the relationships among them.

2.7 Appendix: Technical points needed to complete the proof of Proposition 2.3

We outline three technical points in the proof of Proposition 2.3.

First, suppose that D is not of the form $\log n / \log k$. One can still choose a sequence of approximate decompositions of X into n_i rescaled copies, each reduced by a factor of k_i, with the property that

$$\log n_i / \log k_i \to D, \tag{2.45}$$

and carefully modify the limits in the above computations. The problem and its solution are reminiscent of Euclid's treatment of incommensurable quantities in discussing similar triangles, and illustrate the need for suitable generalizations of the scaling dimension.

Secondly, we have not considered whether the coverings obtained by rescaling are minimal. One needs to check that the computations hold if all coverings are replaced by minimal coverings (in terms of the number of open sets). This is tedious and conveys little insight; the details can be found in the references cited above.

Finally, the limit $\lim_{m \to \infty}$ in formula (2.34) is equivalent to the limit $\lim_{r \to 0}$ for a *restricted set of values of r*, namely the sequence $\{r_0/k^m\}$. This limit must be replaced by the more general limit $\lim_{r \to 0}$. The basic idea, in the simplest case, is that, given any r, we can choose m so that

$$mr_0/k^{m+1} < r \leqslant mr_0/k^m, \tag{2.46}$$

simply by letting m be the integer part of the solution t (greatest integer less than or equal to t) to the equation $r = r_0/k^t$, that is,

$$m = \text{int}[\log(r_0/r)/\log k]. \tag{2.47}$$

The inequality (2.46) implies that

$$N(r_0/k^m) \leqslant N(r) < N(r_0/k^{m+1}) = kN(r_0/k^m), \tag{2.48}$$

since reducing the size of the open balls in a covering can never decrease the number needed to cover a given set. It is now straightforward to check that

$$-\log N(r_0/k^m)/\log(r_0/k^{m+1}) < -\log N(r)/\log r$$
$$< -\log N(r_0/k^{m+1})/\log(r_0/k^m). \tag{2.49}$$

But the inequality (2.49) can be rewritten as

$$-\log N(r_0/k^m)/[\log(r_0/k^m) - \log k] < -\log N(r)/\log r$$

$$< -[\log N(r_0/k^m) + \log k]/\log(r_0/k^m),$$

$$(2.50)$$

which traps the term $-\log N(r)/\log r$ in between two terms with the same limit. Thus the limit as $r \to 0$ can be computed in terms of the sequence $\{r_0/k^m\}$ as $m \to \infty$.   □

3

Dimension of patterns

3.1 Introduction

In the last chapter we introduced two related concepts of dimension for fractals: the scaling dimension (Section 2.3) and the Hausdorff dimension (Section 2.5). We then described the axioms for fractal processes, whose graphs form prototype random fractals (Section 2.4), and power laws, the fundamental algebraic property that links dimension and self-similarity (Section 2.6). We shall now develop alternative equivalent definitions of dimension, with practical methods for computing the dimension of random fractals, in particular, the graph of Brownian motion. We shall also develop some additional properties of dimension.

We shall find and use two key principles of scaling behaviour. The first, which we call the 'telescope–microscope' principle, states that reducing the scale of measurement (for example, the size of open sets used in computing the Hausdorff dimension or the size of a figure in measuring the scaling dimension) of an object X by a factor s is equivalent to scaling X up by a factor s. The second principle states that any measurement the 'mass' of a fractal of Hausdorff dimension D contained in a box of side s must scale as s^D. This provides alternative measurements of D.

There are many natural ways to determine the mass of a fractal set X and thus its Hausdorff dimension. One of the simplest is to lay grids of several scales over the object and count the number of squares in each grid which meet the object. This leads to the 'box dimension' (Section 3.2). The box dimension and the scaling dimension are complementary approaches to the same problem of measurement: shrinking the boxes is equivalent to magnifying the object. The box dimension has been used to compute the dimension of fractal curves such as perimeters of islands and boundaries of leaves (Morse *et al.* 1985). Our discussion is similar to that of McGuire (1991). An alternative for discrete sets is to count the number of points within typical grid squares. The grid squares are replaced by balls around points of the object leading to the cluster or correlation (Hentschel and Procaccia 1983) dimension (Section 3.3). Lovejoy *et al.* (1986) used the cluster dimension to characterize the clustering of weather stations. Hastings *et al.* (1992) used the cluster dimension to study patterns of pancreatic islets. The cluster dimension is also used to find patterns in time series (Grassberger and Procaccia 1983; Sugihara and May 1990a); see Chapter 7. We next develop (in Sections 3.4 and 3.5) two special techniques for computing the dimension of fractal curves and boundaries. The first, the dividers method (Richardson 1961; see Mandelbrot 1977, 1982; Sugihara and May 1990a), arises from measuring the 'length' of a fractal curve, such as a coastline or river, using a range of scales. The second method uses a power law relationship between areas and perimeters of a family of fractal islands. Applications include Lovejoy's (1982) study of rainfall and cloud patterns.

Korcak's (1938) patchiness exponent for the distribution of areas of a family of islands, a scaling exponent closely related to the fractal dimension, introduced in Chapter 2, is developed in Section 3.6.

Although a particular approach may be needed to compute the dimension in any given application, all of these approaches measure essentially the same quantity, which thus deserves to be called the *fractal dimension*.

We conclude this chapter with a description of the main mathematical properties of dimension functions in Section 3.7. Chapter 4 covers additional methods for computing the dimension of graphs of functions and introduces Fourier transform techniques. Details of the Fourier transform are given in Chapter 5.

3.2 The box dimension

The scaling dimension of the square is based on representing the square as an almost-disjoint union of small unit squares. The Hausdorff dimension of an object is based on covering the object by small disks or balls. These ideas converge in familiar area formulae such as the formula for the area of a circle,

$$A = \pi r^2. \tag{3.1}$$

Formula (3.1) arises from approximating the circle with an almost-disjoint union (of approximately πr^2) small unit squares, and taking a suitable limit as the unit squares themselves are shrunk. Although the circle is not self-similar—it cannot be written as an almost-disjoint union of smaller circles—the formula for its area is a simple power law. The exponent in this power law is the familiar Euclidean dimension of the circle. A similar, larger circle of radius kr will cover k^2 times as many small unit squares as a circle of radius r. Equivalently, shrinking the unit squares by a factor of k will yield a smaller scale (and *relatively* larger circle), again requiring k times as many unit squares (see Fig. 3.1).

The box dimension of a subset X of the plane is defined similarly, by counting the number of small unit boxes which intersect X. (This is the simplest way to handle partially occupied boxes.) Let $N(\Delta s)$ denote the number of boxes in a grid of linear scale Δs which meet X. Then X has box

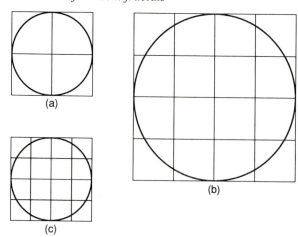

Fig. 3.1 Scaling of circle. (a) A small circle meets 4 unit squares. (b) If the radius of the circle is doubled, then it meets 16 of the same unit squares. (c) Alternatively, the unit squares in (a) are reduced by a linear factor of $\frac{1}{2}$. The same circle as (a) now meets 16 unit squares. Part (c) may be obtained by reducing both the circle and the unit squares in (b) by a linear factor of $\frac{1}{2}$. We call this the telescope–microscope principle.

dimension D if $N(\Delta s)$ satisfies the power law

$$N(\Delta s) \approx c(1/\Delta s)^D \qquad (3.2)$$

asymptotically in the sense that

$$\lim_{\Delta s \to 0} N(\Delta s)\, \Delta s^D = c. \qquad (3.3)$$

The box dimension D is computed by solving equation (3.2) asymptotically for D, obtaining

$$D = \lim_{\Delta s \to 0} \left[-\log N(\Delta s)/\log \Delta s\right]. \qquad (3.4)$$

This is the same as the formula for the Hausdorff dimension (2.28), except that minimal covers by 'round' open balls (which are hard to count) are replaced with covers by boxes in a grid (which are easy to count). Consequently the Hausdorff dimension equals the box dimension whenever the latter is defined (see Proposition 3.4 below).

As in computations of the Hausdorff dimension and scaling dimension in Chapter 2, the limit in formula (3.4) need only be computed for a sequence of scales $\{\Delta s_i\}$ which approaches 0 (see the Appendix to Chapter 2). Figure 3.2 illustrates the computation of the box dimension of the standard Cantor

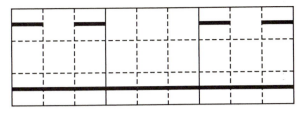

Fig. 3.2 Box dimension of the Cantor set and the line. The Cantor set, as well as the stage of the Cantor set shown above, meets 2 of the larger unit boxes (bounded by solid lines) and 4 of the smaller unit boxes (bounded by dashed lines). The smaller boxes are $\frac{1}{3}$ the linear scale of the larger boxes. Continuing this calculation over all scales yields a box dimension of log 2/log 3, or about 0.63 for the Cantor set. The line segment shown above meets 3 of the larger unit boxes and 9 of the smaller unit boxes. Continuing this calculation over all scales yields a box dimension of 1. Compare Fig. 2.6 on the topological dimension.

set. The box dimension D, whenever it exists, is well defined because of the following 'critical property'. More generally, all such exponents defined by power laws share similar critical properties.

PROPOSITON 3.1 Suppose that for a given exponent D, and for sufficiently small Δs, the set $\{N(\Delta s)\,\Delta s^D\}$ is bounded above and is bounded below by a strictly positive number. Then

$$\lim_{\Delta s \to 0} N\,\Delta s^{D'} = \infty \quad \text{for } D' < D \qquad \text{and} \qquad \lim_{\Delta s \to 0} N\,\Delta s^{D'} = 0 \quad \text{for } D' > D,$$

(3.5)

and thus the limit $\lim_{\Delta s \to 0} N\,\Delta s^{D'}$ is only both finite and nonzero at the single value $D' = D$.

Proof. First suppose that $D' < D$. Then $N\,\Delta s^{D'} = N\,\Delta s^D\,\Delta s^{D'-D}$. As Δs approaches 0, $\Delta s^{D'-D} = 1/\Delta s^{D-D'}$ approaches ∞, and $N\,\Delta s^D$ is bounded strictly above 0 (the weaker condition that $N\,\Delta s^D > 0$ does not suffice). Letting Δs approach 0 yields the first part of equation (3.5). The second part can be shown similarly, yielding the conclusion. □

REMARKS 3.1 The above proposition holds whenever $N(\Delta s)\,\Delta s^D$ has a finite positive limit as Δs approaches 0.

A square of side s has box dimension 2 and two-dimensional measure (area) $A = s^2$. It is also easy to see that the box dimension can be defined for objects in the line, or three-dimensional Euclidean space, or in fact, any Euclidean space and that the box dimension of a set, provided that it is defined, does not depend upon the ambient Euclidean space. However, the situation for the associated D-dimensional measures is more complex, and the box dimension can only be used to compute the measure of subsets of codimension 0 (that is, whose dimension is the same as the dimension of the ambient Euclidean space). Otherwise the more general Hausdorff dimension is needed to define the D-dimensional measure.

PROPOSITION 3.2 Suppose that a figure X has scaling dimension $D = \log n/\log k$. Then the box dimension of X is defined and also equal to D.

Sketch of proof. Suppose that, for a given Δs, c boxes of side Δs are required to cover X. By hypothesis, we may write

$$X = \bigcup_{1 \leqslant i \leqslant n} (1/k)X + a_i, \tag{3.6}$$

an almost-disjoint union of n copies of X, each contracted by a factor k. A single reduced copy of X, not translated, $(1/k)X$, can clearly be covered by c reduced boxes of side $\Delta s/k$. Therefore, except for small discrepancies due to incommensurability of the translations a_i and the unit $\Delta s/k$, X is covered by cn reduced boxes of side $\Delta s/k$.

This argument can be iterated, implying that X is covered by cn^m small boxes, each of side $\Delta s/k^m$. This implies that, as m approaches ∞, the number of boxes N of side $\Delta s/k^m$ required to cover X satisfies

$$\lim_{m \to \infty} N(\Delta s/k^m)D = \lim_{m \to \infty} cn^m(\Delta s/k^m)^D$$

$$= \lim_{m \to \infty} c\Delta s^D(n/k^D)^m$$

$$= \lim_{m \to \infty} c\Delta s^D(n/k^{\log n/\log k})^m \quad \text{(by hypothesis)}$$

$$= \lim_{m \to \infty} c\Delta s^D(n/n)^m$$

$$= c\Delta s^D. \tag{3.7}$$

By proof of the critical property of scaling exponents (Proposition 3.1), formula (3.7) holds for at most one scaling exponent D. As in Section 2.5 and the Appendix to Chapter 2, the above formula holds for all sufficiently small boxes. The conclusion follows. $\square$

3.2.1 *The box dimension of the graph of Brownian motion*

As in Section 2.4, consider the graph of $y = B(t)$ on the unit interval $0 \leqslant t \leqslant 1$ and as an almost-disjoint union of 4 segments, each defined on one-quarter of the unit interval. Suppose that the original graph can be covered by N boxes of a given side Δs. Then each of the 4 segments can be covered by N *rectangular* boxes of width $\Delta s/4$ and height $\Delta s/2$. Each of these rectangular boxes is an almost-disjoint union of two square boxes of side $\Delta s/4$ (see Fig. 3.3). In Brownian motion, $\Delta y^2/\Delta t$ has a finite limit as Δt approaches 0, and therefore the 'approximate slope' (slope of an approximating random walk) $\Delta y/\Delta t$ approaches ∞. Thus, for small Δs, the graph of Brownian motion would have almost vertical slope and meet almost all of the $8N$ boxes of side $\Delta s/4$ in Fig. 3.3. There is nothing special about the number 4: it can be replaced by any perfect square. Extending this argument to all sufficiently small scales yields a box dimension of $\log 8/\log 4 = 1.5$.

The number of pieces in the above subdivision of the t-axis was chosen for convenience. One could instead have divided the t-axis into k^2 pieces for any integer $k > 1$, and covered each of the k^2 segments by N rectangular boxes of width s/k^2 and height s/k. Each of these boxes could then be further subdivided into k square boxes of side s/k^2. As above, most of the time that the graph meets one of the k^2N rectangular boxes of width s/k^2 and height s/k, it will pass through all of the k smaller square boxes it contains. Thus the graph meets k^3N small square boxes of side s/k^2, again yielding $D = \log k^3/\log k^2 = 1.5$. This result is readily extended to generalized Brownian or fractal processes.

PROPOSITION 3.3 *The graph of a fractal process of Hurst exponent H (see Definition 2.3) has box dimension* $2 - H$.

Sketch of proof. First suppose that H is a rational number, say

$$H = m/n. \tag{3.8}$$

One can proceed as in the computation of the box dimension of the graph of Brownian motion, except that the t-axis is subdivided into 2^n segments, the first cover uses rectangular boxes of width $s/2^n$ and height $s/2^m$, and the finer cover uses small square boxes of side $s/2^n$. If the original graph was covered by N boxes of side s, it will also be covered by 2^nN rectangular boxes of width $s/2^n$ and height $s/2^m$, and $2^{2n-m}N$ small square boxes of side $s/2^n$. This yields

$$D = \log 2^{2n-m}/\log 2^n = 2 - m/n. \tag{3.9}$$

Again, the number 2 can be replaced by any integer greater than 1. This argument can be extended to irrational Hurst exponents by following the outline in the first paragraph of the Appendix to Chapter 2. □

REMARKS 3.2 The box dimension of a self-affine figure is sometimes called an elliptic dimension (in contrast to the scaling dimension) in order to reflect the nonisotropic scale changes used in the above calculation (see Schertzer and Lovejoy 1991).

We have seen that the box dimension is easy to compute. The next proposition shows that it agrees with the more general Hausdorff dimension.

(a)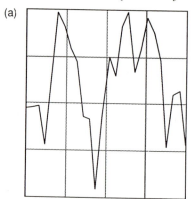

(b)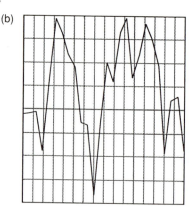

(c)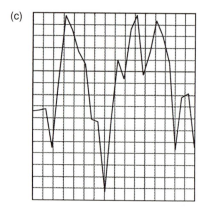

Fig. 3.3   The box dimension of the graph of Brownian motion $y = B(t)$, (a) The graph of $y = B(t)$ with a superimposed grid of *unit* square boxes, each of side Δs. The graph meets 12 of these boxes. (b) The grid is modified by subdividing each unit square box in (a) into 8 rectangular boxes, each of width $\Delta s/4$ and height $\Delta s/2$, corresponding to the self-affine property of Brownian motion. The rescaled function $y = \frac{1}{2}B(4t)$ is statistically similar to the original function $y = B(t)$. Rescaling the graph is equivalent to shrinking the boxes in the covering grid (see the *telescope–microscope principle* of Fig. 3.1). Moreover, the rescaled graph on the interval $0 \leqslant t \leqslant 1$ is equivalent to the original graph on the larger interval $0 \leqslant t \leqslant 4$. Therefore, the graph of $y = B(t)$ can be expected to meet 48 of the rectangular boxes. It actually meets 47 such boxes. (c) The grid in (b) is further modified by subdividing each rectangular box into two smaller square boxes, each of side $\Delta s/4$. Since Brownian motion has a steep 'slope' at such small scales, the graph of $y = B(t)$ can be expected to meet 94 of these small squares, corresponding to the 47 rectangles above. The graph actually meets 90 of these small squares, yielding an apparent box dimension at this scale of $\log(90/12)/\log 4 = 1.4$. The actual value is 1.5.

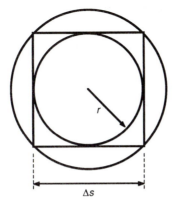

Fig. 3.4 Coverings by square boxes and round balls. A circle of radius r (the smaller circle above) can be covered by a square of side $2r$. Similarly, a square of side Δs can be covered by a circle of radius $\sqrt{2}\,\Delta s/2$ (the larger circle above). Here the ambient dimension $E = 2$.

PROPOSITION 3.4 Let X be a set of Hausdorff dimension D embedded in Euclidean space. Then X also has box dimension D.

Sketch of proof. Define an intermediate covering dimension D' analogous to the Hausdorff dimension, except that square boxes replace open balls. It is not hard to see that if X can be covered by $N(r)$ open balls of radius r then it can be covered by $N(r)$ square boxes of side $2r$ (each circumscribed about an open ball). Similarly, if X can be covered by $N'(\Delta s)$ open square boxes of side Δs, then it can be covered by $N'(\Delta s)$ open balls of radius $E^{1/2}\Delta s/2$ (each circumscribed about an open box), where E is the dimension of the ambient Euclidean space (see Fig. 3.4).

One can easily check that

$$D' = \lim_{u \to 0} \left[-\log N'(u)/\log u \right] = \lim_{u \to \infty} \left[-\log N(u)/\log u \right] = D. \quad (3.10)$$

Now consider covering X by boxes of side Δs from a selected grid, and suppose that N'' such boxes are required. (We shall ignore here the distinction between open and closed boxes. This technical point is readily handled (see Hurewicz and Wallman 1941).) Clearly, except for this point,

$$N'(\Delta s) \leqslant N''(\Delta s). \quad (3.11)$$

But it is easy to see that each box in the more general covering can be covered by at most 2^E boxes from the grid (see also Fig. 3.4) and thus

$$N''(\Delta s) \leqslant 2^E N'(\Delta s). \quad (3.12)$$

Formulae (3.11) and (3.12) imply that

$$D' = \lim_{u \to 0} \left[-\log N'(u)/\log u \right] = \lim_{u \to 0} \left[-\log N''(u)/\log u \right] = D \quad (3.13)$$

and thus the box dimension is D, as required.   □

In conclusion, there is a close relationship between the box dimension and the scaling dimension of self-similar figures. The scaling dimension and box dimension are in fact complementary ways of looking at scaling behaviour. The scaling dimension considers the effect of shrinking or magnifying a figure and the box dimension considers the effect of shrinking or magnifying the underlying grid. We shall define a discrete analogue of the box dimension, the cluster or correlation dimension, in the next section.

3.3 The cluster dimension

Consider now an object X consisting of finitely many points or pixels. Although X has box dimension 0, in many cases a dimension reflecting clustering of the points of X can be defined by power law scaling of the number of intermediate-sized boxes which meet X. For example, let X_1 consist of all the lattice points (points with integral coordinates (m, n)) within some large box with sides S. Consider covering X_1 with boxes of side Δs where

$$1 \ll \Delta s \ll S. \quad (3.14)$$

Since the large box can be divided up into $(S/\Delta s)^2 = S^2/\Delta s^2$ small boxes of side Δs, and most of these boxes contain points in X_1 if $1 < \Delta s < S$, it follows that the box dimension of X_1 in the *scaling range* $1 < \Delta s < S$ is equal to 2. For another example, let X_2 consist of all of the lattice points on the x-axis in some interval $0 \leqslant x \leqslant S$. Since the interval can be covered by $S/\Delta s$ small boxes of side Δs, and most of these boxes contain points in X_2 if $1 < \Delta s < S$, then the box dimension of X_2 in the *scaling range* $1 < \Delta s < S$ is equal to 1.

We could count the number of points within a typical occupied box instead of counting the number of occupied boxes. Within the scaling range $1 < \Delta s < S$, a typical occupied box of side Δs will contain Δs^2 points of X_1, and Δs points of X_2.

We now define the cluster dimension of an object X, motivated by this example. We shall consider occupied boxes of side

$$s = 2r \quad (3.15)$$

centred on points of X. If the average number of points within such a box scales as

$$\text{const} \times s^D \quad (3.16)$$

for some exponent D, then X has cluster dimension D. (Compare Ripley's (1981, p. 150ff.) discussion of Poisson distributions in D-dimensional space.)

Since we only need the exponent in the power law (3.16), the cluster dimension can be defined with disks instead of boxes, and, alternatively, the cluster dimension can be defined with boxes chosen from a specific grid. See the proof that the Hausdorff dimension equals the box dimension (Proposition 3.4).

The first observation, that the cluster dimension can be computed with disks instead of boxes, is useful in applications and in computation since it is both natural and easy to study scaling of the number of points within a distance r of a typical point in a finite point set. Alternatively, one can compute all distances between pairs of points and count the number N of such distances which are less than r. Since the number N' of pairs of points within a distance r of a typical point is $2N$ divided by the number of points in X, N and N' must satisfy the same power law. Thus both methods yield the same dimension. A program for computing the cluster dimension is given in Section 12.6.3.

The second observation relates the cluster dimension to the box dimension. Consider a finite set X of p points in Euclidean space with box dimension D over an appropriate scaling range $s_0 \leqslant s \leqslant s_1$. We shall compute the box dimension using boxes from the grids used to compute the scaling dimension. Suppose that the average occupied box of side s contains k of these pixels and that b boxes are occupied. Clearly,

$$p = kb, \quad \text{and} \quad k = p/b. \qquad (3.17)$$

By scale invariance, the p pixels must be divided among p/k boxes of 'radius' s. Moreover, the number of occupied boxes, b, scales as $(1/s)^D$:

$$b = \text{const} \times (1/s)^D \quad (s_0 \leqslant s \leqslant s_1). \qquad (3.18)$$

Combining the above two formulae yields

$$k = p/[\text{const} \times (1/s)^D] = \text{const} \times s^D \qquad (3.19)$$

Thus the box dimension given by the exponent in the above formula is also equal to D.

We have shown that the box dimension and cluster dimensions are equal, and thus, by Proposition 3.2, both dimensions equal the scaling dimension whenever the latter is defined.

COROLLARY 3.1 Consider a set X of box dimension D. Suppose that A is a finite subset consisting of randomly chosen points from X. Then the cluster dimension of A is equal to D.

Sketch of proof. Suppose that A consists of n points and X meets k boxes. Then the average number of points in each of these boxes is n/k. Now mimic the proof that the cluster dimension equals box dimension (above) considering all boxes which meet X. □

Applications (a) This corollary was used by Hastings *et al.* (1992) to compute the dimension of ductules in the pancreas (see Section 8.3).

(b) Lovejoy *et al.* (1986) used the cluster dimension to estimate the dimension of the global weather sensing network, and then used the corollary and properties of dimension to determine whether strongly clustered storms could be located. Other applications include our observations about connections between neuronal processes (Section 8.4) and about counting species (Section 10.7.6).

(c) Estimates of the dimension of stars and galaxies (Mandelbrot 1977, 1982; Szalay and Schramm 1985) use the cluster dimension at the opposite end of size scales.

(d) The cluster dimension was used by Grassberger and Procaccia (1983) to compute the dimension of strange attractors (see Chapter 7).

3.4 Dimension of boundaries

We now describe additional methods for computing the dimension of the boundary of a fractal island or a collection of fractal islands. Of course, one could apply the general methods for computing the fractal dimension of a set of Euclidean space; for example, one could compute the box dimension of the boundary. However, there are two more specialized natural methods: the dividers method, analogous to the box dimension, and the area–perimeter exponent, analogous to the scaling dimension (Mandelbrot 1977, 1982).

The dividers method is best illustrated by asking a classic question: 'How long is the coastline of Britain?' (Mandelbrot 1977, 1982; cf. Sugihara and May 1990a, which we paraphrase below). This question can be answered fancifully by using a giant to walk around the coast of Britain, counting his steps, or more practically using an appropriate pair of dividers to traverse a map image of the coastline with a polygonal path with steps of length Δs. Let $N(\Delta s)$ denote the number of steps of length Δs required to traverse the coastline.

If the coastline were a simple smooth curve of length L, then the limit

$$\lim_{\Delta s \to 0} N(\Delta s)\Delta s = L \qquad (3.20)$$

is *finite*, and, in fact, formula (3.20) is the formula from elementary calculus for the length of a rectifiable curve. However, empirically, the limit (3.20) is infinite in the case of the coastline of Britain.

Now suppose that the boundary were a fractal curve of scaling dimension $D = \log n/\log k$. Then the boundary would be (at least statistically) an almost-disjoint union of n copies of itself, each reduced by a scale factor of k. Suppose that measurement of the boundary required $N(\Delta s)$ steps of size Δs, and yielded an apparent length of $N(\Delta s)\Delta s$. Consider measurement of the boundary using steps of size $\Delta s/k$. We shall measure each reduced copy directly. Measurement of a reduced copy using steps of size $\Delta s/k$ is equivalent to measurement of the boundary itself using steps of size Δs. Therefore, on average, $N(\Delta s)$ steps of size $\Delta s/k$ will be required to traverse each reduced copy of the boundary, and $nN(\Delta s)$ such steps will be required to traverse the entire boundary. We have shown that

$$N(\Delta s/k) = nN(\Delta s) \tag{3.21}$$

and thus

$$
\begin{aligned}
N(\Delta s/k)(\Delta s/k)^D &= nN(\Delta s)(\Delta s/k)^D \\
&= (nk^D)N(\Delta s)\Delta s^D \\
&= (n/k^{\log n/\log k})N(\Delta s)\Delta s^D \\
&= N(\Delta s)\Delta s^D.
\end{aligned}
\tag{3.22}
$$

If we let k approach ∞, and suitably rescale n using the scaling dimension, Euclid's commensurability argument (cf. Section 2.5 and the Appendix to Chapter 2) implies that

$$\lim_{\Delta s \to \infty} N(\Delta s)\Delta s^D = L_D, \tag{3.23}$$

with L_D nonzero and finite. Thus this process, *the dividers method*, has the scaling dimension D as critical exponent. The dimension D can be found by rewriting formula (3.23) in asymptotic form

$$N(\Delta s) \approx L_D\Delta s^{-D} \quad \text{as} \quad \Delta s \to 0, \tag{3.24}$$

and fitting the data to formula (3.24). The constant L_D in the above formulae is the corresponding D-dimensional measure of the size of the boundary.

More generally, a curve which can be traced with N steps of size Δs can be covered by N disks of radius Δs, and, conversely, a curve which can be covered by N disks of radius Δs can be traced with $2N$ steps of size Δs. Thus the dividers method is just another implementation of the familiar Hausdorff or box dimension and yields the same value. Similar arguments were used to relate the box dimension to the scaling dimension (Proposition 3.2), and the cluster dimension to the box dimension (Corollary 3.1).

In essence, using dividers of step Δs ignores smaller features of the boundary. This is most readily seen by considering the iterative construction of the Koch snowflake (Fig. 2.2). Recall that the beginning stage, stage 0, is an equilateral triangle. Suppose that the perimeter of this triangle is L,

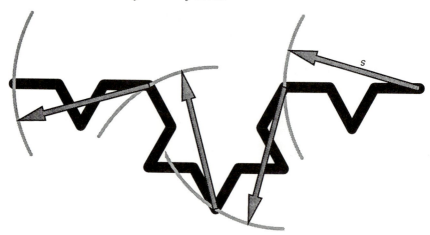

Fig. 3.5   Illustration of the dividers method. Tracing the Koch snowflake with steps of length Δs ignores bumps on scales smaller than Δs.

making each side $L/3$. Then stage n consists of 3×4^n segments, each of length $L/3^{n+1}$. As shown below, tracing the Koch snowflake with dividers of step

$$\Delta s = L/3^{n+1}, \tag{3.25}$$

starting at one of the vertices of the original equilateral triangle, recovers stage n of the triangle, and thus ignores subsequent stages. Even starting at a randomly chosen point in the Koch triangle with a random Δs, not of the form 3^{n+1}, ignores features significantly smaller than Δs, and essentially recovers stage n, where n is given approximately by formula (3.25).

Applications. Bradbury *et al.* (1984) used the dividers method to compute the dimension of boundaries of features in an Australian coral reef (see Chapter 10).

3.5   The area–perimeter exponent

The ideas behind the dividers method yield a related method, using the area–perimeter exponent, for a fractal pattern of many islands of different areas. For such a fractal, scale-invariance implies a power law relation between the area A and perimeter p of each island:

$$p = \text{const} \times A^E. \tag{3.26}$$

We shall see that the exponent E is half the fractal dimension of the boundary as computed by any of the above methods.

Regular objects provide a simple example. For example, for a circle, the perimeter or circumference C, the radius r, and the area A are related by the formulae $C = 2\pi r$ and $A = \pi r^2$.

Thus $r = (A/\pi)^{1/2}$, and

$$C = (2\pi^{1/2})A^{1/2} \tag{3.27}$$

The area–perimeter exponent, $\frac{1}{2}$, is half the fractal dimension, 1, of the smooth boundary of the circle.

We now perform a thought-experiment on more general fractal patterns of islands. Consider for example, two islands of areas A_1 and A_2, and perimeters p_1 and p_2, respectively. We assume, without loss of generality, that the second island has larger area. Define a linear scale factor k by the formula

$$k = A_2^{1/2}/A_1^{1/2}, \quad \text{or equivalently } A_2 = k^2 A_1. \tag{3.28}$$

The square root in formula (3.28) transforms two-dimensional measure into one-dimensional measure. Then the second island is statistically a copy of the first island, enlarged by the (one-dimensional) scale factor k.

Suppose that the perimeters of both islands are measured on the scale Δs, and that N_i steps are required to traverse the ith island. By scale-invariance, measurement of the second island on the scale Δs is equivalent to measurement of the first island on the scale $\Delta s/k$. Since the dimension of the boundary of the first island can be computed by the dividers method,

$$N_2(\Delta s/k)^D = N_1 \Delta s^D, \tag{3.29}$$

an analogue with formula (3.22). Thus

$$N_2 = N_1 k^D = N_1(A_2/A_1)^{D/2} \tag{3.30}$$

by equation (3.29) followed by equation (3.28). Since the perimeters are simply $N_1 \Delta s$ and $N_2 \Delta s$, respectively, formula (3.30) implies that

$$p_2/p_1 = (A_2/A_1)^{D/2}, \tag{3.31}$$

yielding an area–perimeter exponent of $D/2$ for these two islands.

This sketch may be readily formalized to prove that the area–perimeter exponent in general is given by $D/2$, where D is the dimension of the boundary.

A similar thought-experiment relates the area–perimeter exponent to the box dimension. Suppose that the perimeters of a family of islands all have box dimension D. This means that if we approximate the perimeter of these islands using a grid of scale Δs, the number of boxes the perimeter of each island will meet is asymptotically proportional to $(1/\Delta s)^D$. However, as above, shrinking the grid by a factor k is statistically equivalent to magnifying the islands by the same factor k. Shrinking the grid by a factor k asymptotically multiplies the number of boxes met by the boundary by k^D. Thus magnifying each island by a factor k also asymptotically multiplies the number of boxes of fixed scale Δs which meet by the boundary by k^D. In addition, the magnification asymptotically multiplies the area of the island (which is proportional to the number of boxes of any fixed scale met by the island itself) by k^2 since the islands have box dimension 2. However, in a scale-invariant pattern, large islands are statistically similar to magnified small islands. Thus the area and perimeter of various islands, both measured on the same fixed scale, are related by the formula

$$\text{perimeter} \approx \text{const} \times k^D \approx \text{const} \times \text{area}^{D/2}, \tag{3.32}$$

yielding an area–perimeter exponent of $D/2$.

Applications. Krummel *et al.* (1987) (see Chapter 10 below) used the area–perimeter exponent to discuss landscape patterns. Lovejoy's (1982) early work on cloud shapes and rainfall patterns used both the area–perimeter exponent and the cluster dimension.

3.6 Cumulative frequencies: the Korcak patchiness exponent B

Fractal patterns of islands can also be described by another scaling exponent, the exponent B introduced by Korcak (1938, see Chapter 2) to study the distribution of the areas of the islands. It has been conventional to study the cumulative frequency distribution, that is, the number $N(a)$ of islands of area greater than or equal to a. By scale-invariance,

$$N(a) = \text{const} \times a^{-B}. \tag{3.33}$$

The exponent in equation (3.33) must be negative since $N(a)$ is clearly a nonnegative nonincreasing function of a. We shall call B the Korcak patchiness exponent.

Applications. Korcak (1938), see Mandelbrot (1977, 1982), used the exponent B to parametrize the distribution of areas of Aegean islands. Hastings *et al.* (1982) used this method to measure patchiness in vegetative ecosystems. Chapter 10 summarizes the work of Hastings *et al.* (1982), Meltzer (1991), and others in this area.

The exponents B and D are related by simple formulae involving the dimension of the ambient space (Mandelbrot 1977, 1982, p. 118). For fractal sets of islands in the plane

$$B = \tfrac{1}{2}D \quad (0 \leqslant B \leqslant 1), \tag{3.34}$$

and more generally for fractal sets of islands in the n-dimensional Euclidean

Table 3.1 Computation of B and D for the standard Cantor set

Δs	Stage of Cantor set construction (Fig. 2.1)	At scale Δs	
		Number of intervals of length Δs, $n_D(\Delta s)$	Number of gaps of length at least Δs, $n_B(\Delta s)$
1/3	1	2	1
1/9	2	4	$1 + 2 = 3$
1/27	3	8	$1 + 2 + 4 = 7$
1/81	4	16	$1 + 2 + 4 + 8 = 15$
$1/3^n$	n	2^n	$1 + 2 + \cdots + 2^{n-1} = 2^n - 1$

space

$$B = (1/n)D \quad (0 \leqslant B \leqslant 1). \tag{3.35}$$

We call formulae (3.34) and (3.35) the *Mandelbrot formulae*, and sketch a proof of (3.35), largely following Mandelbrot (1977, 1982, p. 118). First, for motivation, consider the process of computing B and D for the Cantor set (Fig. 2.1) C by choosing a sequence of scales Δs, and counting both the number of intervals of length Δs required to cover C, and the number of gaps of length at least Δs between points in C. The results are summarized in the Table 3.1.

The dimension D is the limit

$$\lim_{\Delta s \to 0} [\log n_D(\Delta s)/\log(1/\Delta s)] = \log 2^n/\log 3^n$$

$$= (n \log 2)/(n \log 3)$$

$$= \log 2/\log 3. \tag{3.36}$$

The Korcak exponent B is the limit

$$\lim_{\Delta s \to 0} [\log n_B(\Delta s)/\log(1/\Delta s)] = \lim_{\Delta s \to 0} [\log(2^n - 1)/\log 3^n]$$

$$= \log 2/\log 3, \tag{3.37}$$

since clearly

$$\lim_{\Delta s \to 0} [\log(2^n - 1)/\log 2^n] = 1. \tag{3.38}$$

Thus $D = B$.

REMARKS 3.3 We have restricted the values of Δs in the above limits to values of the form $1/3^n$. A standard argument (see equations 2.46–2.50) above) implies that the above limits are well defined over *all* scales. We sketch the proof. Given a more general scale Δs, choose n such that

$$1/3^{n+1} < \Delta s \leqslant 1/3^n, \tag{3.39}$$

which requires that

$$n = \text{int}(-\log \Delta s/\log 3), \tag{3.40}$$

where $\text{int}(u)$ represents the integer part of u, that is, the greatest integer less than or equal to u. Formula (3.39) implies both that

$$\lim_{\Delta s \to 0} (\log \Delta s)/n = \log 2 \tag{3.41}$$

and, since the number of intervals $n_D(\Delta s)$ of length Δs required to cover the Cantor set is a nonincreasing function of Δs, also that

$$\lim_{\Delta s \to 0} [\log n_D(\Delta s)]/n = \log 2. \tag{3.42}$$

Similarly,

$$\lim_{\Delta s \to 0} [\log n_B(\Delta s)]/n = \log 2 \tag{3.43}$$

Informally, Δs is asymptotic to 3^n, and both $n_D(\Delta s)$ and $n_B(\Delta s)$ are asymptotic to 2^n. Thus the limits in formulae (3.36) and (3.37) are well defined, and $B = D$.

3.6.1 *The general case* (following Mandelbrot 1982, p. 118)

More generally, suppose that a fractal X embedded in the line $\mathbb{R}$ has scaling dimension $D = \log k/\log n$. Suppose that there are $n_B(\Delta s)$ gaps between points of length at least Δs. Then the fractal X is statistically the union of k rescaled copies of itself, each reduced by a factor of n. Each reduced copy contains $n_B(\Delta s)$ gaps between points of length at least $\Delta s/n$. Thus the fractal X asymptotically has

$$n_B(\Delta s/n) \approx k n_B(\Delta s) \tag{3.44}$$

gaps of length at least $\Delta s/n$ (neglecting the order-of-k gaps induced in the process of taking the union). Iterating formula (3.44) implies that

$$n_B(\Delta s/n^m) = k^m n_B(\Delta s) \quad (m = 1, 2, 3, ...) \tag{3.45}$$

and thus that

$$n_B(\Delta s) = \text{const} \times \Delta s^{\log k/\log n} \tag{3.46}$$

As above, formula (3.46) implies that $B = D$. A similar argument shows that

$$D = nB \quad \text{or} \quad B = (1/n)D \tag{3.47}$$

for fractal sets of islands embedded in n-dimensional Euclidean space.

3.7 Properties of dimension

Here is a summary of the main properties of dimension functions, including all of the equivalent definitions of fractal dimension D as well as the topological dimension D_{top}. Consider first objects X and Y of fractal dimension $D(X)$ and $D(Y)$, respectively, embedded in a Euclidean space E.

3.7.1 *Subset relationships*

Suppose first that X is a subset of Y. Then the fractal dimension D and the topological dimension D_{top} both satisfy the subset relationships

$$D(X) \leqslant D(Y) \quad \text{and} \quad D_{\text{top}}(X) \leqslant D_{\text{top}}(Y).$$

3.7.2 *Topological dimension and fractal dimension*

These are related as follows. For any topological manifold X or for the Euclidean space itself, the fractal dimension equals the topological dimension. The class of topological manifolds includes the regular objects of Euclidean geometry (points, line segments, arcs, polygons), the graphs of smooth (differentiable) functions, and open sets and their topological closures. Loosely, the 'nice' objects of calculus are topological manifolds. In general we have

$$D_{\text{top}}(X) \leqslant D(X) \leqslant D(E) = D_{\text{top}}(E).$$

We may therefore interpret the difference $D(X) - D_{\text{top}}(X)$ as a measurement of the irregularity or nonmanifold behaviour of X. We similarly interpret the relationship between $D(X)$ and $D(E) = D_{\text{top}}(E)$ as measuring the extent to which X locally fills up E.

3.7.3 *Intersection relationships*

One may see quite easily that two randomly drawn lines in the Euclidean plane intersect in a point, and that two randomly drawn planes in Euclidean 3-space intersect in a line. These are consequences of the fact that, for two subspaces X and Y in general position (a precise statement of random

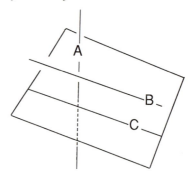

Fig. 3.6 Intersection relationships. Line A and horizontal plane P are in general position in three-dimensional space, and they therefore intersect in a set of dimension $1 + 2 - 3 = 0$, that is, a point. Line B parallel to plane P and line C contained within plane P are not in general position. Line B does not meet P; line C meets P in a line.

position) in Euclidean space E, we have

$$D_{top}(X \cap Y) = D_{top}(X) + D_{top}(Y) - D_{top}(E)$$

and

$$D(X \cap Y) = D(X) + D(Y) - D(E).$$

Application (see Mandelbrot 1977, 1982; Hastings *et al.* 1982; Vicsek 1989; Erzan and Sinha 1991). Consider a *self-similar* fractal pattern in space-time, where space, as represented by the vector x is n-dimensional. This makes the (x, t)-space $(n + 1)$-dimensional, with the $t =$ constant hyperplanes n-dimensional and the $x =$ constant lines one-dimensional. By the intersection relationships, the $t =$ constant sections have typical scaling dimension

$$D_{t = const} = D + n - (n + 1) = D - 1,$$

and the $x =$ constant sections have typical scaling dimension

$$D_{x = const} = D + 1 - (n + 1) = D - n.$$

Thus

$$D_{t = const} = D_{x = const} + n - 1.$$

Erzan and Sinha (1991) use these results to study the dynamics of the Bak–Tang–Weisenfeld (1987) model for earthquakes. *Caution*: these results do not hold for more general self-affine fractals.

3.7.4 *Projection relationships*

Consider a projection from Euclidean 3-space to the Euclidean plane. The entire Euclidean 3-space as well as most planes in 3-space project to the

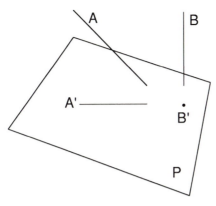

Fig. 3.7 Projection relationships. Lines A and B are projected vertically (ortho-gonally) onto horizontal plane P, yielding the line A′ (the usual case—line A is in general position), and the point B′ (since the line B is vertical, and hence not in general position).

entire Euclidean plane (only planes parallel to a 'vertical' line in Fig. 3.7 project to lines). Most lines project to lines, and all points to points. Note that we might consider a random plane (with respect to the direction of projection) instead of considering most planes. These geometric projection relationships reflect the fact that for a scale-invariant object X embedded in Euclidean m-space, its projection to Euclidean n-space has fractal dimension $\max\{D(X), n\}$, and similarly for topological dimension.

Moreover, the same result holds for random maps of maximum rank for Euclidean m-space to Euclidean n-space. A map has maximum rank if its Jacobian matrix has maximum rank (n in this case). This implies that the map locally is a projection onto Euclidean n-space, but the direction of the projection may change from point to point in Euclidean m-space.

4

Dimension of graphs of functions

4.1 Introduction

In the previous chapter we developed the theory and computation of the fractal dimension (and related exponents) of spatial patterns. We now consider the dimension of graphs of functions, and especially the graphs of time series. Although the methods of Chapter 3 can be applied to sets of *fractal islands* of the form

$$\{t : f(t) \geqslant c\}, \tag{4.1}$$

there are additional methods for computing the fractal dimension (and related exponents) of graphs of scale-invariant or self-affine functions and time series (regarded as functions with discrete domains) which use more of the information available.

We shall develop these methods in the next three sections: Section 4.2, second moment techniques which compute the fractal Hurst exponent by fitting the data to the axiomatic description of a fractal process (see also Definition 2.3); Section 4.3, closely related local second moment (correlation-like) techniques (see also formula (2.26)); and Section 4.4, growth of range techniques which make use of the renormalization property of fractal processes (see also formula (2.27)).

The computational methods of the previous chapter also yield scaling exponents for the graphs of fractal processes; see Section 4.5. Section 4.6 extends Section 3.6.1 by describing Mandelbrot (1977, 1982) formulae for these new exponents. Fourier transform techniques are introduced in Section 4.7, and are developed more fully in Chapter 5. We conclude the introduction with one crucial warning: one must be careful wtih fractal analysis of short time series just like any other statistical technique. One easy way to estimate confidence limits is to delete an initial or final segment of the series and repeat the analysis. Simulation methods can also be used.

Finally, the study of the fractal exponents of time series and functions poses two special questions: whether to detrend the data and when to take sums. The first question concerns data which may contain linear or cyclic trends. The second question is best illustrated by an example. In modelling river discharges, Hurst (1951, 1956) and Mandelbrot (1977, 1982) considered the cumulative discharge (the sum of discharges to date) rather than the discharges themselves as a fractal process. The rationale for this choice, and, more generally, answers to these questions, will be given in Chapter 9.

4.2 Second moment techniques

The third axiom for fractal processes (see Definition 2.3, restated below) provides a natural way to determine the Hurst exponent H in a postulated fractal model of experimental data. For a fractal process and any Δt, the corresponding increments Δy have expectation 0 and satisfy the equation

$$E(\Delta y^2) = c \, \Delta t^{2H}. \tag{4.2}$$

For an example from a fractal process, the mean value of Δy^2 is an unbiased estimator of expectation $E(\Delta y^2)$. The exponent $2H$ in equation (4.2) can be determined empirically by using linear regression to fit log-transformed data to the log-transformed version of equation (4.2):

$$\log E(\Delta y^2) = \log c + 2H \log \Delta t. \tag{4.3}$$

In addition, the exponent $2H$ can be determined locally in the scaling interval from Δt to $2 \, \Delta t$ by computing the ratio

$$E([y(t + 2 \, \Delta t) - y(t)]^2)/E([y(t + \Delta t) - y(t)]^2). \tag{4.4}$$

More precisely, axiom (iii) implies that

$$H = \tfrac{1}{2}\{\log E([y(t + 2 \, \Delta t) - y(t)]^2) - \log E([y(t + \Delta t) - y(t)]^2)\}, \tag{4.5}$$

Since the Hurst exponent H of a fractal process is independent of the time step, local computations of the form (4.5) can be used to test whether a given process is fractal. One can substitute more general intervals, or even windows containing more than two data points, to obtain similar local fractal exponents. Similar windowing techniques can be applied to all computations of fractal exponents. These techniques will be used later to determine scaling regions in ecosystem patterns (Section 10.5) and in natural time series (Chapters 9 and 11).

Caution and critique. In the case of long time series, the mean increment Δy will be very small compared with the second moments discussed above, and the terms second moment and variance may be used interchangeably. This is, however, not the case for short time series, and we prefer to test the hypotheses about the expectation and second moment independently. A similar caution holds for correlation techniques below.

4.3 Local second moment (correlation-like) techniques

The Hurst exponent can also be determined from the coefficient of correlation between successive increments (axiom (iv) of Definition 2.3) using formula (2.26). Under the assumption that the increments Δy have expectation 0, the coefficient of correlation ρ is determined directly from its definition:

$$\rho = \frac{E([y(t + 2\,\Delta t) - y(t + \Delta t)][y(t + \Delta t) - y(t)])}{\{E((y(t + 2\,\Delta t) - y(t + \Delta t))^2)E([y(t + \Delta t) - y(t)]^2)\}^{1/2}}, \quad (4.6)$$

the expected value of the product of successive increments divided by the geometric mean of their second moments. The mean values of the terms $[y(t + 2\,\Delta t) - y(t + \Delta t)][y(t + \Delta t) - y(t)]$, $[y(t + 2\,\Delta t) - y(t + \Delta t)]^2$, and $[y(t + \Delta t) - y(t)]^2$ serve as unbiased estimators for the expectations. The local Hurst exponent is then given by the formula

$$2^{2H} = 2 + 2\rho, \quad \text{or } H = \log(2 + 2\rho)/\log 4, \quad (4.7)$$

as derived in formula (2.26).

More generally, formulae (4.6) and (4.7) compute the local fractal exponent over the scaling interval from Δt to $2\,\Delta t$, even when the expectation is not zero, and ρ is *not* the coefficient of correlation. The axioms for a fractal process may be tested by repeating these local computations for several values of Δt. This method will be used in Chapter 11 to test whether population fluctuations are in fact fractal.

4.3.1 *Thought-experiment*

Consider a short time series obtained from a fractal process with $H > \frac{1}{2}$, meaning that the increments are expected to be positively correlated. In this case we expect the mean increment to be nonzero. Can the above techniques still be used? Yes, if the coefficient ρ is used to estimate the local fractal exponent, and is not interpreted as a coefficient of correlation.

4.4 Growth of range

The concept of renormalization may be applied to compute the fractal exponent H from the growth of range, with one caveat stated below. This method is closely related to the rescaled range of Mandelbrot and Van Ness (1968) and Mandelbrot and Wallis (1969).

The range of a fractal process $\{y(t)\}$ over a time interval Δt is defined to be the difference between the maximum and minimum values of $y(t)$ in that interval. We shall let $R(\Delta t)$ denote the average range of the process $\{y(t)\}$ over all intervals of duration Δt. The scaling behaviour of the range $R(\Delta t)$ as a function of the duration Δt can be easily determined by renormalization.

If $\{y(t)\}$ is a fractal process with fractal exponent H, then, for any constant $c > 0$, the process

$$y_c = (1/c^H)y(ct) \qquad (4.8)$$

is another fractal process with the same statistics. Thus the processes $\{y(t)\}$ and $\{y_c(t)\} = \{(1/c^H)y(ct)\}$ should have the same expected range, over all intervals. Formula (4.8) would then imply that the range of the process $\{y_c(t)\}$ over an interval of duration Δt is the $1/c^H$ times the range of the process $\{y(t)\}$ over an interval of duration $\Delta t/c$. Replacing $\Delta t/c$ by Δt implies that the range should scale as

$$R(\Delta t) = c\,\Delta t^H. \qquad (4.9)$$

However, in the case of a fractal process in discrete time, formula (4.9) only holds for sufficiently long time intervals, since there are not enough data points within a short time interval to adequately determine the range.

Unfortunately, many natural processes can only be sampled in discrete time. Therefore, formula (4.9) must be corrected before it can be used to determine the fractal exponent of real data. It is clear that restricting the computation of the range of $y(t)$ over a time interval by using just a few points in that interval underestimates the range, and thus for short intervals, $R(\Delta t)$, will grow faster than Δt^H (see Table 4.1).

Thus the process defined by sampling Brownian motion at discrete times has $H = 0.63$, *not* the expected 0.5. This investigation was motivated by studying time series of bird populations with relatively few data points (see Chapter 11, especially Section 11.6).

Table 4.1 Growth of range $R(\Delta t)$ for Δt at most 5, compared with the value $(\Delta t)^{1/2}$ expected from the axioms for a continuous-time Brownian process

Time lag, Δt	Expected $R(\Delta t)$ from analysis of all paths	$(\Delta t)^{1/2}$
1	1.0	1.0
2	1.5	1.414
3	2.0	1.732
4	2.375	2.0
5	2.75	2.236

4.5 Other techniques

The techniques of Chapter 3 may also be applied to the determination of scaling exponents for the graphs of fractal processes. We shall recall the most useful techniques here.

The fractal dimension D of the graph may be determined with the box dimension or a similar technique. One can also determine the fractal dimension D_0 of the set of zero-crossings of the graph. More formally, D_0 is just the fractal dimension of the set

$$\{t : y(t) = 0\}. \tag{4.10}$$

Since the set of zero-crossings is just the intersection of the graph with the line $y = 0$ in two-dimensional Euclidean space,

$$D_0 = D + 1 - 2 = D - 1. \tag{4.11}$$

One can also compute the Korcak patchiness exponent B for the intervals on the t-axis where $y(t)$ is positive, or even for all of the intervals which result from cutting the t-axis at zero-crossings. The exponent B is defined by the power law

$$N(L > l) = cl^{-B}, \tag{4.12}$$

where $N(L > l)$ denotes the number of intervals of length greater than l. Then

$$B = D_0, \tag{4.13}$$

following an argument similar to those in Section 3.6.1.

The calculation of the fractal dimension of the graph of Brownian motion implies that

$$D = 2 - H, \tag{4.14}$$

and thus

$$D_0 = 1 - H \quad \text{and} \quad B = 1 - H. \tag{4.15}$$

This completes the list of Mandelbrot relations among the exponents.

These relations have important consequences for applications, for example, the Korcak exponent for the distribution of durations of floods or droughts is closely related to the power law for the range of river discharges (see Chapter 9).

4.6 Fourier transform techniques

The use of Fourier transform techniques provides another extremely useful exponent H. We shall briefly motivate Fourier transforms here, but defer most of the details to Chapter 5.

Consider first the idea of a surface, such as the surface of the ocean, made up of waves upon waves. We can build up such surfaces mathematically by adding up a sequence of simpler waves, defined by sine and cosine functions. Fourier transform techniques represent the surface in terms of the amplitudes (and signs, that is, Fourier coefficients) of the sine and cosine waves required to build up the surface. If the surface is fractal, then the amplitudes should satisfy a power law which depends upon the fractal exponents of the surface. In the simpler case of a fractal curve, the power spectrum (set of variances of the Fourier coefficients) scales as

$$c/f^{1+2H}, \tag{4.16}$$

where f is the frequency. Surfaces may be resolved in two directions. Power spectra of surfaces scale as

$$c/(f_1^{1+2H}f_2^{1+2H}) = c/(f_1 f_2)^{1+2H}, \tag{4.17}$$

where f_1 and f_2 are the frequencies in the two directions. Although these methods are readily and frequently used to construct fractal curves and surfaces (the Mandelbrot–Weierstrass fractals of Berry and Lewis (1980)), some care is needed in their application (see especially the caution in Section 5.1 below).

The Fourier transform

5.1 Introduction

Chapter 4 introduced the use of Fourier transform techniques in the calculation of fractal exponents. This chapter describes the mathematical details. The Fourier transform of a time series shows how that series is built up from simple periodic functions. Similarly, the Fourier transform of spatial data describes spatial periodicity. We shall show that the Fourier coefficients associated with Brownian and fractal processes satisfy power law scaling rules corresponding to the scaling rules which characterize their self-similar behaviour. The scaling rules of the Fourier coefficients are easily used to calculate the corresponding fractal exponents. Conversely, the most common class of random fractal curves and surfaces, the Mandelbrot–Weierstrass fractals (Berry and Lewis 1980; Mandelbrot 1982), is constructed by applying the inverse Fourier transform to appropriately scaled 'Fourier coefficients' with random phases. This construction builds up complex patterns from waves upon waves, recalling an intuitive picture of the surface of the ocean.

   This chapter is organized as follows. Section 5.2 reviews the foundations of Fourier transforms, beginning with the Fourier series of a continuous function f defined on the unit interval $0 \leqslant t \leqslant 1$. The Fourier series represents the function f as a sum of sine and cosine functions. Section 5.3 develops practical implementations. The Fourier transform is usually implemented using a computer algorithm called the fast Fourier transform (FFT) (see Aho *et al.* 1974; Burrus and Parks 1985; Horowitz and Sahni 1978; Kreyszig 1988). The FFT uses complex exponentials in place of the equivalent sine and cosine terms. The set of Fourier coefficients of a function is frequently called its *spectrum*. The fast Fourier transform has a long history (see Cooley *et al.* 1967; Aho *et al.* 1974, p. 276) extending back to Runge and König (1924), Danielson and Lanczos (1942), and Good (1958). The most common implementation is due to Cooley and Tukey (1965). The basic terminology of power spectra (essentially squares of amplitudes of the spectrum) is reviewed in Section 5.4. In Section 5.5, we derive the Fourier series of Brownian motion from the axioms, and state the corresponding scaling rule for both Brownian and fractal processes. The converse process for constructing Mandelbrot–Weierstrass fractals (Berry and Lewis 1980) is given in Section 5.6.

Caution and critique. Although the Fourier transform of any sequence of Fourier coefficients (spectrum) is periodic, that is, $f(0) = f(1)$, Fourier analysis is frequently applied to nonperiodic functions such as random walks. The lack of periodicity will generate additional high-frequency terms in any discrete Fourier transform, and therefore scaling regions *must* be chosen carefully to avoid such high-frequency terms.

5.2 Review of Fourier transforms

The Fourier transform of a function f describes f as a sum of multiples of simple periodic functions, sines, and cosines, at a fundamental frequency and its harmonics. The coefficients of these functions describe the behaviour of f at scales corresponding to their respective frequencies. Both periodicity and random-walk behaviour may be clearly represented in terms of corresponding patterns in the Fourier transform of f.

5.2.1 *The standard case:* the Fourier transform of a continuous function f on the closed interval $[0, 1]$

Let f be a continuous function defined on the closed interval $[0, 1]$ which satisfies the condition $f(0) = f(1)$. The Fourier transform in this case represents the function f as a sum (its Fourier series) of simple periodic functions:

$$f(t) = \sum a_n \cos 2\pi nt + \tfrac{1}{2}a_0 + \sum b_n \sin 2\pi nt. \tag{5.1}$$

Here all sums run over $n = 1, 2, 3, \dots$, and the coefficients are given by the integrals

$$a_n = 2 \int f(t) \cos 2\pi nt \, dx \quad (n = 0, 1, 2, 3, \dots) \tag{5.2}$$

and

$$b_n = 2 \int f(t) \sin 2\pi nt \, dt \quad (n = 1, 2, 3, \dots), \tag{5.3}$$

where all integrals are over the unit interval.

Assuming the decomposition (5.1), the formulae (5.2) and (5.3) follow from orthogonality relations among the sine and cosine functions:

$$
\left.
\begin{aligned}
\int \sin 2\pi mt \sin 2\pi nt \, dt &= \begin{cases} 0 & \text{if } m \neq n, \\ \tfrac{1}{2} & \text{if } m = n, \end{cases} \\[2mm]
\int \cos 2\pi mt \cos 2\pi nt \, dt &= \begin{cases} 0 & \text{if } m \neq n, \\ 1 & \text{if } m = n = 0, \\ \tfrac{1}{2} & \text{if } m = n > 0, \end{cases} \\[2mm]
\int \sin 2\pi mt \cos 2\pi nt \, dt &= 0.
\end{aligned}
\right\} \tag{5.4}
$$

The orthogonality relationships are easily derived using integration formulae from elementary calculus. Formula (5.3) follows from integrating

$$\int f(t) \sin 2\pi mt \, dt = \int f(t) \left(\sum a_n \cos 2\pi nt + \tfrac{1}{2}a_0 + \sum b_n \sin 2\pi nt \right) dt$$

$$= \sum a_n \int \cos 2\pi nt \sin 2\pi mt \, dt + \tfrac{1}{2}a_0 \int \sin 2\pi mt \, dt$$

$$+ \sum b_n \int \sin 2\pi nt \sin 2\pi mt \, dt = \tfrac{1}{2}b_m. \tag{5.5}$$

Formula (5.2) uses a similar integration of $\int f(t) \cos 2\pi mt \, dt$.

We shall need one key formal property of the Fourier transform, namely *linearity*. The Fourier transform can be considered as a function T from the vector space of all suitable (piecewise continuous) functions on the interval $0 \leqslant t \leqslant 1$ to the vector space of all sequences of Fourier coefficients. Let c be a constant and write the Fourier transform of a function f as $T(f)$. Then

$$T(f_1 + f_2) = T(f_1) + T(f_2) \quad \text{and} \quad T(cf) = cT(f). \tag{5.6}$$

This property, linearity, follows immediately from the analogous property for the integrals (5.2) and (5.3) which define the Fourier coefficients.

5.3 Implementing the Fourier transform

The efficient application of Fourier transform techniques relies upon an extremely clever and efficient numerical algorithm for computing transforms. This algorithm is known as the fast Fourier transform (FFT), and is available on most mathematical application packages. Although we shall not develop the FFT algorithm itself because of its mathematical complexity, we shall indicate its relation to the Fourier transform of Section 5.2, above. A program is included in Chapter 12.

The FFT implements a discrete version of the Fourier transform called the discrete Fourier transform (DFT). The DFT requires only a finite sequence of data points, for example from a time series, in place of the continuous function f above, and is thus more useful than the standard Fourier transform in most applications. The DFT and FFT are most easily described with complex exponential functions in place of sine and cosine functions.

The relationship between sine and cosine functions and complex exponential functions is given by the formulae

$$\exp(i\phi) = \cos \phi + i \sin \phi \tag{5.7}$$

and correspondingly

$$\cos \phi = [\exp(i\phi) + \exp(-i\phi)]/2 \quad \text{and} \quad \sin \phi = [\exp(i\phi) - \exp(-i\phi)]/2i.$$
(5.8)

These relationships can be used to rewrite the Fourier series (5.1) in terms of complex exponentials and complex coefficients by replacing the building blocks $\cos 2\pi nt$ and $\sin 2\pi nt$ by corresponding combinations of $\exp(2\pi int)$ and $\exp(-2\pi int)$. This yields

$$f(t) = \sum c_n \exp(2\pi int),$$
(5.9)

where the summation extends over all integers $n = \ldots, -2, -1, 0, 1, 2, \ldots$.

Formulae for the coefficients c_n may be derived by making similar substitutions into formulae (5.2) and (5.3), or by directly using the orthogonality relationships

$$\int \exp(2\pi imt) \exp(-2\pi int) \, dt = \begin{cases} 1 & \text{if } m = n, \\ 0 & \text{if } m \neq n. \end{cases}$$
(5.10)

The integral is again over the unit interval. Note the use of the complex conjugate $\exp(-2\pi int)$ in place of $\exp(2\pi int)$, here and below, in integrals of certain products. We obtain the formula

$$c_n = \int f(t) \exp(-2\pi int) \, dt$$
(5.11a)

Caveat. Some authors (Aho *et al.* 1974; Horowitz and Sahni 1984) define the discrete Fourier transform as the complex conjugate of our transform. For example, they replace formula (5.11a) by

$$c_n = \int f(t) \exp(2\pi int) \, dt,$$
(5.11b)

and make a corresponding change in formula (5.7). Others (Burrus and Parks 1985; Kreyszig 1988) follow our sign convention. Since we shall only need the magnitudes of the Fourier coefficients in order to compute fractal exponents, the choice of sign convention has no effect outside of the mathematical derivation.

The discrete Fourier transform (DFT) is defined by constructing a discrete analogue of the standard Fourier transform. This avoids errors implicit in discrete numerical integration problems with extending a time series to a continuous function. The function f on the unit interval is replaced by the vector

$$\mathbf{f} = (f_0, f_1, \ldots, f_{n-1}).$$
(5.12)

The DFT represents f as a sum of vector analogues of complex exponential functions, namely the vectors v_j whose kth entry is

$$\exp(2\pi ijk/n). \tag{5.13}$$

For example, the vector $v_0 = (1, 1, \dots, 1)$ is the analogue of the function $\exp(2\pi i0t)$, the vector

$$v_1 = (1, \exp(2\pi i/n), \exp(2\pi i \cdot 2/n), \exp(2\pi i \cdot 3/n), \dots, \exp(2\pi i \cdot (n-1)/n))$$

is the analogue of the function $\exp(2\pi it)$, and the vector

$$v_j = (1, \exp(2\pi ij/n), \exp(2\pi ij \cdot 2/n), \exp(2\pi ij \cdot 3/n), \dots, \exp(2\pi i \cdot j(n-1)/n))$$

is the analogue of the function $\exp(2\pi ijt)$.

We shall use the methods of linear algebra in order to obtain the DFT. Let M be the matrix whose jth column is just v_j. Then the (j, k)th entry of M (jth column, kth row) is given by

$$M_{jk} = \exp(2\pi ijk/n). \tag{5.14}$$

For example, if $n = 4$, then

$$M = \begin{bmatrix} 1 & 1 & 1 & 1 \\ 1 & i & -1 & -i \\ 1 & -1 & 1 & -1 \\ 1 & -i & -1 & i \end{bmatrix} \tag{5.15}$$

Let N be the conjugate of M:

$$N_{jk} = \exp(-2\pi ijk/n). \tag{5.16}$$

By direct calculation, the product MN is the diagonal matrix whose diagonal entries are n. Therefore,

$$(1/n)MN = I, \tag{5.17}$$

the identity matrix, that is, the matrices $(1/n)M$ and $(1/n)N$ are inverses, and the columns of M are a basis for the vector space of n-dimensional complex-valued vectors.

Since M and N are symmetric matrices, we may interchange rows and columns almost at will. In particular, we shall identify f with its transpose, the column vector f^T, whenever f is multiplied by a matrix, and simply write f in place of f^T. Now write

$$f = (1/n)MNf \tag{5.18}$$

and consider the transformations

$$f \to Nf \tag{5.19}$$

and

$$g \to (1/n)Mg \tag{5.20}$$

The transformation (5.19) transforms f into the vector $g = Nf$ whose entries g_j are the coefficients in the expansion of f as a linear combination of the rows of the matrix $(1/n)M$:

$$f = \sum_j g_j(1/n)v_j, \tag{5.21}$$

as required. Therefore the transformation (5.19) is the discrete Fourier transform or DFT, and (5.20) is the inverse discrete Fourier transform.

We note that the DFT and inverse DFT are automatically linear because multiplication by a matrix is linear, see (5.6). Conversely, one could derive the matrix representation formally from linearity. We chose the more computational and less abstract approach.

Moreover, there is no problem with convergence or the way with which vectors can be represented by Fourier transform methods because there are no limits or infinite sums. One can show that there is a close relationship between the Fourier transform of a function and the DFT of the time series obtained by sampling the function at discrete intervals. The details are complex and are omitted.

We now consider a computer implementation of the DFT: $f \to Mf$. Computing the product Mf is straightforward, but involves a large computational cost for large n. Computing the product of an $n \times n$ matrix with a vector of length n in the usual way requires n^2 (complex) multiplications and n^2 (complex) additions, thus the order of n^2 operations. (This is usually denoted $\mathcal{O}(n^2)$ operations in the computational complexity literature.) For large n the computational cost may be prohibitive. For example, if $n = 10\,000$ in a time series problem, then there are the order of 10^8 operations. The analogous two-dimensional DFT for two-dimensional arrays (functions of two discrete variables) is even more expensive: a small two-dimensional (300×300) DFT requires the order of 10^{10} operations.

However, whenever n is a power of two, the matrices M used to compute the DFT have useful symmetry properties. These significantly speed up the calculation, and ultimately lead to the fast Fourier transform (FFT). These properties can be used to reduce the computation of a DFT of length n to two computations of a DFT of length $n/2$, together with n additions and subtractions. Figure 5.1 illustrates this symmetry in the case $n = 4$.

At the cost of increasing the program complexity, the FFT uses these symmetry properties to sharply reduce the number of computations needed

The DFT of a vector of length 4

$$
\begin{bmatrix}
1 & 1 & 1 & 1 \\
1 & i & -1 & -i \\
1 & -1 & 1 & -1 \\
1 & -i & -1 & i
\end{bmatrix}
\begin{bmatrix}
w \\ x \\ y \\ z
\end{bmatrix}
=
\begin{bmatrix}
w + x + y + z \\
w + ix - y - iz \\
w - x + y - z \\
w - ix - y + iz
\end{bmatrix}
=
\begin{bmatrix}
(w + y) + (x + z) \\
(w - y) + i(x - z) \\
(w + y) - (x + z) \\
(w - y) - i(x - z)
\end{bmatrix}
$$

can be computed in terms of two DFTs of length 2

$$
\begin{bmatrix}
1 & 1 \\
1 & -1
\end{bmatrix}
\begin{bmatrix}
w \\ y
\end{bmatrix}
=
\begin{bmatrix}
w + y \\
w - y
\end{bmatrix}
$$

$$
\begin{bmatrix}
1 & 1 \\
1 & -1
\end{bmatrix}
\begin{bmatrix}
x \\ z
\end{bmatrix}
=
\begin{bmatrix}
x + z \\
x - z
\end{bmatrix}
$$

and the additions and subtractions

$$
(w + y) \pm (x + z) \quad \text{and} \quad (w - y) \pm i(x - z).
$$

Fig. 5.1 Symmetry properties of DFT matrices.

to compute the DFT. The DFT requires the order of $n \log_2 n$ operations for a vector of size n. This represents a significant practical improvement over the computational cost of the DFT (the order of n^2 operations) for $n > 16$. Figure 5.2 illustrates the relative computational costs of the FFT and the basic DFT algorithms.

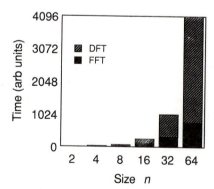

Fig. 5.2 Relative computational cost of the discrete Fourier transform (DFT) and the fast Fourier transform (FFT).

5.4 Power spectra

It is convenient to introduce several alternative representations of the spectra, or coefficients in the Fourier transform. These coefficients are frequently described in terms of amplitudes and phases. The power spectrum of a function is the sequence of squares of the amplitudes of its Fourier coefficients. We review the precise definitions here.

First, let f be a continuous function on the interval $[0, 1]$ with $f(0) = f(1)$, as in Section 5.2, and recall the Fourier series of f:

$$f(t) = \sum a_n \cos 2\pi nt + \tfrac{1}{2}a_0 + \sum b_n \sin 2\pi nt. \qquad (5.22)$$

We combine cosine and sine terms of the same frequency, obtaining the equation

$$f(t) = \tfrac{1}{2}a_0 + \sum (a_n \cos 2\pi nt + b_n \sin 2\pi nt). \qquad (5.23)$$

In addition, since the formula for the cosine of the difference of two 'angles' yields

$$c_n \cos[2\pi(nt - \delta_n)] = c_n \cos 2\pi\delta_n \cos 2\pi nt + c_n \sin 2\pi\delta_n 2\pi nt, \qquad (5.24)$$

we can rewrite the Fourier series (5.23) as

$$f(t) = \sum c_n \cos[2\pi(nt - \delta_n)], \qquad (5.25)$$

where the *amplitude*

$$c_0 = \tfrac{1}{2}a_0, \qquad c_n = (a_n^2 + b_n^2)^{1/2} \quad (n > 0), \qquad (5.26)$$

and the *phase* δ_n (well defined modulo 1 if $c_n > 0$) satisfies the conditions

$$a_n = c_n \cos \delta_n \quad \text{and} \quad b_n = c_n \sin \delta_n \qquad (n > 0). \qquad (5.27)$$

The use of c_n here should not be confused with a different use in the DFT and FFT.

Formula (5.25) clearly expresses the idea that a continuous function may be built up of waves upon waves, at various scales $1/n$, of appropriate amplitudes c_n and phases δ_n.

It is much easier to describe amplitudes and phases in the case of complex exponential representations of the Fourier transform, and in the case of the DFT and FFT. In both cases, the coefficients c_n contain both the amplitude and phase information. Any complex number $z = x + iy$ can be written as the product of a real number r and a complex number $e^{i\phi}$ of magnitude 1, $e^{i\phi}$, namely $re^{i\phi}$, where the *amplitude*, denoted $|z|$, is given by

$$|z| = r = (x^2 + y^2)^{1/2}$$

and the *phase* ϕ (well defined in the interval $0 \leqslant \phi < 2\pi$ if $r > 0$) satisfies the conditions

$$x = r \cos \phi, \qquad y = r \sin \phi;$$

compare with formula (5.26) and (5.27). This yields the *amplitude* (denoted $|c_n|$) and *phase* for the FFT and DFT (the phase is interpreted slightly differently from that in the standard Fourier transform).

The amplitudes are usually represented by the *power spectrum* which represents the 'energy' associated with each frequency. The power spectrum is just the sequence of squares of amplitudes, although in the discrete case (DFT and FFT), we restrict the frequency j to the interval

$$0 < j < n/2, \tag{5.28}$$

where n is the number of data points (components in the vector f and its transform).

Technical remarks. The coefficient c_0 is essentially the average value of the components of the vector f. For $j > n/2$, the frequencies fold back, in that the frequency associated with the vector

$$v_j = (1, \exp(2\pi i j/n), \exp(2\pi i j \cdot 2/n), \exp(2\pi i j \cdot 3/n), \dots , \exp(2\pi i \cdot j(n-1)/n))$$

$$(n/2 < j < n-1)$$

is $n - j$. However, the corresponding Fourier coefficients provide no new information in the case of the DFT or FFT of a real-valued vector since for real data, the amplitudes $|c_j|$ and phases ϕ_j satisfy the symmetry conditions

$$|c_j| = |c_{n-j}|, \quad \phi_j = -\phi_{n-j} \qquad (0 < j < n/2). \tag{5.29}$$

This follows from an easy calculation (cf. Burrus and Parks 1985, p. 27).

5.5 The Fourier transform of Brownian motion

Since Brownian motion (and its fractal generalization) are scale-invariant, their Fourier transforms should also be scale invariant, and thus (the statistics of) their Fourier coefficients should satisfy appropriate power laws. We shall show how the power law for Brownian motion follows formally and easily from the axioms for Brownian motion and from elementary calculus. These power laws will be used in the next section and in Chapter 12 to generate a class of fractals called Mandelbrot–Weierstrass fractals. The results of this technical section are summarized in the next section.

Let f be a Brownian function on the interval $[0, 1]$ with $f(0) = f(1)$. We shall calculate the spectrum of f, or more precisely, scaling rules for its power spectrum. In doing this calculation, it is easiest to use complex exponentials, and then to calculate statistics on the distribution of the random variables C_n corresponding to the Fourier coefficients

$$c_n = \int f(t)\, e^{-2\pi i n t}\, dt \qquad (5.30)$$

for randomly chosen Brownian functions f. We shall see that the Fourier coefficients have expected value 0, random phases, and variances which satisfy power law scaling.

PROPOSITION 5.1 The expected value $E(C_n)$ is 0.

Proof. The key idea is linearity of the expected value of a random variable. Fix a value of n. We shall compute the expectation

$$E\left(\int f(t) \exp(-2\pi i n t)\, dt\right) \qquad (5.31)$$

from the definition of the Riemann integral. For any Riemann sum which approximates the integral in (5.31), we have

$$E\left(\sum f(t_j^*) \exp(-2 i n t_j)\right) = \sum E(f(t_j^*) \exp(-2\pi i n t_j))$$

$$= \sum \exp(-2\pi i n t_j)\, E(f(t_j^*)) = 0, \qquad (5.32)$$

since, for at any time t_j^*, the expected value $E(f(t_j^*))$ of Brownian motion is 0 (there is no drift toward the right or left). Passing to the Riemann integral by taking limits as the mesh of the Riemann sum approaches 0 implies that the expectation (5.31) is zero, as required. $\square$

In fact, we have shown that the real and imaginary parts of the expectation of the complex random variable

$$C_n = X_n + i Y_n \qquad (5.33)$$

are both zero. We shall use this result later in describing the statistics of the phases of the spectrum.

We now proceed to calculate the scaling behaviour of the expected value of the power spectrum

$$E(|C_n|^2) = E\left(\left|\int f(t)\, e^{-2\pi i n t}\, dt\right|^2\right). \qquad (5.34)$$

(Since the expected value of the coefficient C_n is 0, the expected value of the power spectrum coefficient $E(|C_n|^2)$ is also the variance of the coefficient C_n.) In order to do this, we calculate $E(|C_n|^2)$ as a function of n and $E(|C_1|^2)$.

PROPOSITION 5.2

$$E(|C_n|^2) = (1/n^2)E(|C_1|^2) \quad (n > 0).$$ (5.35)

Proof. Let $n > 0$ below. The first step involves integrating

$$\int f(t) \exp(-2\pi i n t) \, dt$$ (5.36)

by a substitution which in essence replaces the factor $\exp(-2\pi i n t)$ by $\exp(-2\pi i u)$. To do this, let $u = nt$, and thus

$$t = u/n \quad \text{and} \quad dt = du/n.$$ (5.37)

Integration by substitution yields

$$\int_0^1 f(t) \exp(-2\pi i n t) \, dt = (1/n) \int_0^n f(u/n) \exp(-2\pi i u) \, du$$

$$= (1/n)\left(\int_0^1 f(u/n) \exp(-2\pi i u) \, du \right.$$

$$+ \int_1^2 f(u/n) \exp(-2\pi i u) \, du$$

$$\left. + \cdots + \int_{n-1}^n f(u/n) \exp(-2\pi i u) \, du \right).$$ (5.38)

The required expected value may be computed by now computing the square of the magnitude of both sides of equation (5.38).

$$\left| \int_0^1 f(t) \exp(-2\pi i n t) \, dt \right|^2$$

$$= (1/n)^2 \left| \int_0^1 f(u/n) \exp(-2\pi i u) \, du + \int_1^2 f(u/n) \exp(-2\pi i u) \, du \right.$$

$$\left. + \cdots + \int_{n-1}^n f(u/n) \exp(-2\pi i u) \, du \right|^2$$

$$= (1/n)^2 \sum_j \left| \int_{j-1}^j f(u/n) \exp(-2\pi i u) \, du \right|^2$$

$$+ (1/n)^2 \sum_{j \neq k} \left(\int_{j-1}^j f(u/n) \exp(-2\pi i u) \, du \right.$$

$$\left. + \int_{k-1}^k f(u/n) \exp(-2\pi i u) \, du \right).$$ (5.39)

Here the bar denotes the complex conjugate and the last equality requires rewriting the square of each magnitude $|z|^2$ as the product of z and its complex conjugate. However, for $j \neq k$,

$$E\left(\left|\int_{j-1}^{j} f(u/n) \exp(-2\pi iu) \, du \times \int_{k-1}^{k} f(u/n) \exp(-2\pi iu) \, du\right|\right)$$

$$= E\left(\int_{j-1}^{j} f(u/n) \exp(-2\pi iu) \, du \times \overline{\int_{k-1}^{k} f(u/n) \exp(-2\pi iu) \, du}\right)$$

$$= E\left(\int_{j-1}^{j} [f(u/n) - f(j-1)/n] \exp(-2\pi iu) \, du\right.$$

$$\left. \times \overline{\int_{k-1}^{k} [f(u/n) - f(k-1)/n] \exp(-2\pi iu) \, du}\right)$$

(since the Fourier transform is linear and the Fourier series of a constant is just the constant itself, or alternatively by a direct calculation)

$$= E\left(\int_{j-1}^{j} [f(u/n) - f(j-1)/n] \exp(-2\pi iu) \, du\right)$$

$$\times E\left(\overline{\int_{k-1}^{k} [f(u/n) - f(k-1)/n] \exp(-2\pi iu) \, du}\right)$$

(since the two integrands and thus the two integrals are independent)

$$= 0 \tag{5.40}$$

(by Proposition 5.1: the expected value of the Fourier coefficients of Brownian motion is 0).

Therefore equation (5.39) implies

$$\left|\int_{0}^{1} f(t) \exp(-2\pi int) \, dt\right|^2$$

$$= (1/n)^2 \left(\left|\int_{0}^{1} f(u/n) \exp(-2\pi iu) \, du\right|^2 + \left|\int_{1}^{2} f(u/n) \exp(-2\pi iu) \, du\right|^2\right.$$

$$\left. + \cdots + \left|\int_{n-1}^{n} f(u/n) \exp(-2\pi iu) \, du\right|^2\right). \tag{5.41}$$

Moreover, each term on the right-hand side of equation (5.41) has the same expected value, so that we need merely rewrite

$$\int_0^1 f(u/n) \exp(-2\pi i u)\, du \tag{5.42}$$

as an integral of a Brownian function defined directly on the *u-axis*. It is here that we invoke the self-similarity of Brownian functions, namely, that for any nonzero constant c, and any Brownian function f, the rescaled function $c^{-1/2} f(cs)$ is itself a Brownian function with the same statistics as the original function $f(s)$ (see Section 2.4). Therefore, the statistics of the integral (5.42) above are the same as those of the rescaled integral

$$n^{1/2} \int_0^1 f(u/n) \exp(-2\pi i u)\, du. \tag{5.43}$$

Thus

$$E\left(\left| \int_0^1 f(u/n) \exp(-2\pi i u)\, du \right|^2\right) = E\left((1/n)\left| \int_0^1 f(u) \exp(-2\pi i u)\, du \right|^2\right)$$

$$= (1/n)\, E\left(\left| \int_0^1 f(u) \exp(-2\pi i u)\, du \right|^2\right). \tag{5.44}$$

The calculations of formulae (5.42)–(5.44) may be summarized in the statement

$$E\left(\left| \int_{j-1}^j f(u/n) \exp(-2\pi i u)\, du \right|^2\right) = (1/n)\, E\left(\left| \int_0^1 f(u) \exp(-2\pi i t)\, dt \right|^2\right). \tag{5.45}$$

We now apply expected values to both sides of equation (5.41) and use formula (5.45) to conclude that

$$E\left(\left| \int_0^1 f(t) \exp(-2\pi i n t)\, dt \right|^2\right) = (1/n^2)\, E\left(\left| \int_0^1 f(u) \exp(-2\pi i u)\, du \right|^2\right). \tag{5.46}$$

Thus

$$E(|C_n|^2) = (1/n^2)\, E(|C_1|^2), \tag{5.47}$$

as required.   □

A closer look at the details of the above proof, including a careful examination of the choice of a random sample from all Brownian motions, implies the following result.

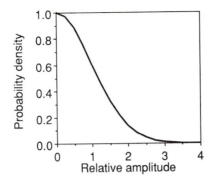

Fig. 5.3 Distribution of the probability density function for the amplitudes of a typical Fourier coefficient c_n.

PROPOSITION 5.3 The distributions of the Fourier coefficients C_n are independent and Gaussian as complex-valued random variables.

This means that the real and complex components are each independent Gaussian random variables, and the amplitudes form 'half' a normal distribution, with probability density function falling off as $\exp(-\text{const} \times |c_n|^2)$. The coefficients of cosine and sine terms of a real representation are each independent and Gaussian.

Finally, we have the following result.

PROPOSITION 5.4 The phases of the Fourier coefficients are uniformly distributed on the interval $[0, 2\pi]$.

Proof. The probability density function for the phase ϕ is given by the following double integral over the sector $\phi < \theta < \phi + \Delta\phi$ up to a constant factor:

$$\iint \exp(-\tfrac{1}{2}x^2) \exp(-\tfrac{1}{2}y^2) \, dx \, dy = \iint \exp[-\tfrac{1}{2}(x^2 + y^2) \, dx \, dy$$

$$= \iint \exp(-\tfrac{1}{2}r^2) \, dx \, dy$$

$$(\text{since } r^2 = x^2 + y^2)$$

$$= \iint \exp(-\tfrac{1}{2}r^2) r \, dr \, d\theta$$

$$= \int_{\phi}^{\phi + \Delta\phi} d\theta = \Delta\phi. \tag{5.48}$$

Since the integral (5.48) is independent of the phase ϕ, the phase has a uniform distribution. (The proof is reminiscent of the elementary calculus trick for computing the definite integral of $\exp(-\tfrac{1}{2}x^2)$ over the real line.)   □

5.6 The Fourier transform of fractal processes and Mandelbrot–Weierstrass fractals

We summarize all of the results in the previous section, and their generalization to fractal processes, in the following theorem.

THEOREM 5.1 Let f be a fractal function on the interval $[0, 1]$ with $f(0) = f(1)$, and with fractal similarity exponent H. Let $\{c_n : n > 0\}$ be the spectrum (Fourier series of c_n). Then each c_n is an independent sample from a complex-valued normal distribution of expected value 0, and expected variance (square of absolute value)

$$\text{const} \times n^{-1-2H}. \tag{5.49}$$

COROLLARY 5.1 The phases are independent and uniformly distributed on the interval $[0, 2\pi]$.

COROLLARY 5.2 The power spectrum of a fractal process satisfies the power law const $\times n^{-1-2H}$.

The above results suggest a technique for generating a class of random fractals called the Mandelbrot–Weierstrass fractals by generating appropriate spectra and then applying the inverse Fourier transform. In order to generate these fractals (Berry and Lewis 1980),

(a) choose a fractal exponent H (H must be in the interval $0 < H < 1$ in order to obtain all of the dimension relationships);

(b) choose the number N of terms (Fourier coefficients) to be used in the simulation (note that N is also the highest frequency involved);

(c) choose N independent random phases $\phi_1, \dots, \phi_N$, uniformly from the interval $[0, 2\pi]$;

(d) choose 'amplitudes' $|c_n|$ from normal distributions with mean 0 and variance proportional to n^{-1-2H} (negative values are acceptable here, or alternatively take the absolute value of the normal variates);

(e) form the spectrum $\{|c_n| \exp(i\phi_n)\}$;

(f) apply the inverse Fourier transform (preferably the inverse FFT) to the above spectrum to obtain a complex-valued fractal process $f(t)$; and, finally,

(g) take the real part of $f(t)$ to obtain the required fractal (since we imposed no symmetry condition on the spectrum).

Note that the inverse FFT is the conjugate of the FFT, up to scale factors, so that the ordinary FFT may be used instead. A computer implementation of the above algorithm is given in Section 12.5.

This algorithm may be readily extended to two dimensions, obtaining Mandelbrot–Weierstrass fractal sheets defined by real-valued functions $z = f(x, y)$. The *two-dimensional* Fourier transform of such a fractal sheet requires *two* integrations, in the x- and y-directions, and thus the amplitudes of the two-dimensional Fourier coefficients scale as

$$c(m, n) = \text{const} \times m^{-\frac{1}{2}-H} n^{-\frac{1}{2}-H}, \tag{5.50}$$

and the corresponding power spectrum as

$$\text{const} \times m^{-1-2H} n^{-1-2H}. \tag{5.51}$$

Fractal islands are readily defined as sets of points (x, y) with $z \geq 0$ in a sea of points with $z < 0$ and bounded by shorelines with $z = 0$ (see Fig. 1.2).

Historical remarks. Weierstrass 'invented' Mandelbrot–Weierstrass fractal curves in the nineteenth century (before the science of fractals) in order to write down examples of continuous nowhere-differentiable curves. For $H < 1$, Mandelbrot–Weierstrass fractals have this property.

Part III

The bridge to applications

The next chapters develop and illustrate the main techniques for the fractal modelling of spatial (and temporal) patterns, beginning with the basic statistics and factors affecting the choice of modelling techniques. The potential breadth of fractal modelling is shown by applications to earthquake models and developmental biology in Chapter 8 and to weather and climate in Chapter 9.

6

Modelling spatial and temporal patterns

6.1 Introduction

The purpose of this chapter is to identify the main techniques in fractal modelling of spatial (and temporal) patterns. Korcak (1938) found a hyperbolic distribution

$$N(a) = \text{const} \times a^{-B} \tag{6.1}$$

for the areas of Aegean islands by using linear regression to fit log-transformed data to the log transform of equation (6.1):

$$\log N(a) = \log(\text{const}) - B \log a. \tag{6.2}$$

These are the first key steps in fractal modelling: the choice of an appropriate power law, the application of log transforms, and finally the use of linear regression to fit a log-transformed linear model.

The organization of this chapter follows these steps, beginning with a summary of fractal exponents in Section 6.2 and a review of linear regression in Section 6.3. Fractal behaviour can be at least partially tested by evaluating the goodness of fit in linear regression: both visually and with the use of the coefficient of correlation. Simulation methods can also be used; see Section 6.4. However, there are other useful tests and important deviations from fractal behaviour. These are discussed in Sections 6.5 and 6.6.

6.2 Summary of fractal exponents

Tables 6.1 and 6.2 summarize the fractal exponents developed in Part II.

It is frequently desirable to use several techniques for computing fractal exponents and then compare the results (see especially Chapter 11 for an illustration).

6.2.1 *Mandelbrot (1977, 1982) formula (see Sections 3.6 and 4.5)*

In many cases, the exponents B, D, and H are related. The Korcak patchiness exponent B of fractal patterns of islands in Euclidean n-space, and the fractal dimension D of their boundaries are related by the formula $D = \min\{nB, n\}$.

Table 6.1   Fractal exponents of spatial patterns

Dimension	Exponent, formula, definition	Typical applications
Scaling dimension	D, $n = \text{const} \times k^D$, if X is a union of n similar copies of itself, each reduced by a linear scaling factor of k	Branching processes such as vascularization, stream order, etc.
Hausdorff dimension	D, $n = \text{const} \times k^D$, where the smallest cover of X by balls of radius $1/k$ consists of n such balls	Mostly theoretical foundations
Box dimension	D, $n = \text{const} \times k^D$, if X intersects n boxes of side $1/k$ formed by a grid in the ambient space E	Computable version of Hausdorff dimension, 'available habitat' in a leaf (Morse 1985), Sections 3.1–3.2
Cluster (correlation dimension)	D, $n = \text{const} \times r^D$, if X is a discrete set and there are n points of X within a radius r of a typical point of X	Dimension of discrete patterns such as pancreatic islets (Hastings *et al.* 1992), Section 8.3 below
Korcak patchiness exponent	B, $n = \text{const} \times a^{-B}$, if B is a union of disjoint 'islands', and n islands have measure (in the ambient space E) at least a	Aegean islands (Korcak 1938), ecosystem patterns (Hastings *et al.* 1982), Chapter 10 below
Hurst exponent	H, $\Delta z = \text{const} \times \Delta s^H$, if X is the graph of a Brownian function f, Δs denotes the length scale in the domain of f, and Δz denotes the length scale in the range of f	Time series (Hurst 1951, 1956; Sugihara and May 1990a), Chapters 9 and 11 below

The 'minimum' is required since $D \leqslant n$ by the subset relationship for dimension, but B may exceed 1 in some applications (Meltzer 1990; Meltzer and Hastings 1992; see Chapter 9 below).

The fractal dimension D of a the graph (in n-dimensional space) of a fractal process with Hurst exponent H on $(n-1)$-dimensional space is $D = n - H$.

Table 6.2 Time series techniques (largely computation of the Hurst exponent H)

Technique	Formula, definition	Typical applications
Second moment (loosely, 'growth of variance')	$\Delta x^2 = \text{const} \times \Delta t^{2H}$, where Δt is the time step, and Δx^2 the second moment of the corresponding spatial increments	Axioms, population fluctuations, Chapter 11 below
Growth of range	$R(\Delta t) = \text{const} \times \Delta t^H$, where Δt is the time step, and $R(\Delta t)$ is the mean range (maximum value of x minus minimum value of x) over time intervals of duration Δt. *Caution*: a correction factor is needed if Δt is small. See Section 4.4	Most common technique for time series (Hurst 1951, 1956; Sugihara and May 1990a), Chapters 9 and 11 below
Local second moment	$2^{2H} = 2 + 2\rho$, where ρ is the correlation between successive spatial increments *in the case* where the increments have zero expectation. See Section 4.3 for the general case	Same as growth of variance, good test for multiscaling, Section 8.2 and Chapters 10 and 11 below
Power spectrum	$E(f) = \text{const} \times f^{-1-2H}$, where f is the frequency, and $E(f)$ the power spectrum coefficient at frequency f	Connection with spectral analysis, Mandelbrot–Weierstrass fractals, simulation, Chapter 5

6.3 Linear regression

Linear regression is used to determine fractal exponents as slopes of log-transformed data. Here are the key ideas (see Hogg and Tanis 1977, pp. 232–7, 324–6; Draper and Smith 1981).

6.3.1 *Notation*

For simplicity, in the remainder of this section, we shall represent *log-transformed* data points by order pairs (x_i, y_i), and discuss the use of *linear regression* to fit such log-transformed experimental data with linear functions

$$y = a + bx. \tag{6.3}$$

Here and below there are n data points (x_i, y_i) and all sums range over $1 \leqslant i \leqslant n$.

If $y = a + bx$, we may associate with each value x_i of the independent variable x, the predicted y-value

$$y_i = a + bx_i \tag{6.4}$$

and the corresponding *residual* or unexplained error

$$y_i - a - bx_i. \tag{6.5}$$

Linear regression minimizes these errors by minimizing the sum of their squares

$$\sum (y_i - a - bx_i)^2, \tag{6.6}$$

or, equivalently, their average

$$\sigma^2 = (1/n) \sum (y_i - a - bx_i)^2. \tag{6.7}$$

For this reason, linear regression is called a *least squares* method.

The parameters a and b (in precise terms, *maximum likelihood estimates of a and b*) are readily determined using the techniques of elementary calculus. The sum of squares of residuals, $\sum (y_i - a - bx_i)^2$, is a quadratic function of the variables a and b, and is therefore readily minimized by taking partial derivatives with respect to a and b, and setting them equal to 0. This yields a pair of linear equations in the variables a and b. Their solution is given by

$$\left.\begin{aligned}
b &= (n \sum x_i y_i - \sum x_i \sum y_i)/(n \sum x_i^2 - (\sum x_i)^2), \\
a &= (\sum y_i - b \sum x_i)/n.
\end{aligned}\right\} \tag{6.8}$$

The formula for the slope b can be written as the quotient

$$b = \text{covar}/\text{var}(x) = \text{covar}/\sigma_x^2, \tag{6.9}$$

where covar denotes the covariance of x and y, and is understood roughly as

$$\Delta x\, \Delta y / \Delta x^2 = \Delta y / \Delta x,$$

the formula for the slope of a line.

There is a simple relationship between the covariance of x and y and the size of the unexplained errors. The *coefficient of correlation* ρ between x and y is defined by

$$\begin{aligned}
\rho &= \text{covar}/[\text{var}(x)\,\text{var}(y)]^{1/2} \\
&= \text{covar}/[\sigma_x^2 \sigma_y^2]^{1/2} \\
&= (n \sum x_i y_i - \sum x_i \sum y_i)/[(n \sum x_i^2 - (\sum x_i)^2)(n \sum y_i^2 - (\sum y_i)^2)]^{1/2},
\end{aligned}$$

$$\tag{6.10}$$

and measures the extent to which changes in the independent variable x are accompanied by corresponding changes in the dependent variable y. Then one can see that the variance σ^2 of unexplained errors (which have mean 0) and the variance σ_y^2 of y are related by a formula involving the coefficient of correlation ρ:

$$\sigma^2 = (1 - \rho^2)\sigma_y^2. \tag{6.11}$$

If the slope b is nonzero, then the coefficient of correlation ρ and the slope b have the same sign; otherwise the coefficient of correlation is necessarily zero.

The distribution of experimental values of a, b, and ρ is readily determined analytically, under the usual assumptions that the independent variables can be measured precisely and that the residuals are independent samples from a normal distribution of fixed variance. (We shall not need to consider the distribution of a.)

Let B denote the random variable corresponding to the slope of a regression line through n points with nominal slope b_0. Then the transformed random variable

$$T = (B - b_0)[(n - 2)\sigma_x^2]^{1/2}/[(1 - \rho^2)\sigma_y^2]^{1/2} \tag{6.12}$$

has a Student's t-distribution with $n - 2$ degrees of freedom. Solving for B yields

$$B = b_0 + T[(1 - \rho^2)\sigma_y^2]^{1/2}/[(n - 2)\sigma_x^2]^{1/2}, \tag{6.13}$$

where T denotes a random variable with the Student's t-distribution. The Student's t-distribution with k degrees of freedom has variance

$$k/(k - 2) \tag{6.14}$$

provided that $k > 2$. Therefore, B follows a similar distribution with mean 0 and variance

$$\text{var}(B) = (1 - \rho^2)\sigma_y^2/(n - 4)\sigma_x^2, \tag{6.15}$$

and standard deviation

$$[\text{var}(B)]^{1/2} = [(1 - \rho^2)\sigma_y^2]^{1/2}/[(n - 2)\sigma_x^2]^{1/2}. \tag{6.16}$$

Confidence limits are readily obtained from tables of the Student's t-distribution.

For n larger than approximately 25 to 30, the normal approximation to the t-distribution implies that T is approximately normal with mean 0 and variance 1. Thus B is approximately normal with mean b_0 and variance

$$\text{var}(b) = (1 - r^2)\sigma_y^2/(n - 2)\sigma_x^2). \tag{6.17}$$

In this case, confidence limits for b_0 are readily obtained using calculators or tables for the normal distribution, for example b_0 lies in the interval

$$b - 1.96[\text{var}(b)]^{1/2} \leqslant b_0 \leqslant b + 1.96[\text{var}(b)]^{1/2}. \qquad (6.18)$$

The statistics of B can be understood by using formula (6.11) for the variance σ^2 of the residual or unexplained errors (differences between predicted and observed y-values) to rewrite the formula for the variance of B. We obtain

$$\text{var}(B) = (1 - \rho^2)\sigma_y^2/(n - 4)\sigma_x^2 = \sigma^2/(n - 4)\sigma_x^2, \qquad (6.19)$$

the quotient of the variance of the residuals by the variance of x and a measure of the amount of averaging implicit in the formula (6.8) for the slope b. These results can be used to *estimate* confidence limits for fractal exponents; see the programs in Chapter 12.

Similarly let R denote the random variable corresponding to the coefficient of correlation in samples of n points with nominal coefficient of correlation ρ_0. If $\rho_0 = 0$, then the transformed random variable

$$T = (n - 2)^{1/2}R/(1 - R^2)^{1/2} \qquad (6.20)$$

also has a Student's t-distribution with $n - 2$ degrees of freedom. More generally, the transformed random variable

$$W = \tfrac{1}{2}\ln\left[(1 + R)/(1 - R)\right] \qquad (6.21)$$

is approximately normal with mean

$$\tfrac{1}{2}\ln\left[(1 + \rho_0)/(1 - \rho_0)\right] \qquad (6.22)$$

and variance

$$1/(n - 3). \qquad (6.23)$$

6.4 Simulation methods

These methods for determining confidence limits are readily supplemented by simulation methods. Moreover, it is probably best to use simulation methods to develop appropriate confidence limits because of the difficulty in accounting for all errors. For example, experimental determination of the cluster dimension D of a discrete fractal set X underestimates the actual value of D. The cluster dimension is determined experimentally by counting the number $n(r)$ of points within a radius r of each point of X, and fitting the data to a hyperbolic distribution

$$n(r) = \text{const} \times r^D. \tag{6.24}$$

For large r, many of the disks $B(x, r)$ of radius r go outside the boundary and contain fewer points than expected. Thus the range of r must be suitably restricted. Even so, a computer simulation gave an experimental $D = 1.91 \pm 0.24$ for 100 replicates, each consisting of 20 points randomly distributed in the plane (see Section 8.3 below). Simulation has many other applications; see, for example, Section 4.4 on the Hurst exponent for short time series.

However, confidence limits for fractal exponents can be readily obtained with simulation methods. For example, assuming a given hypothesis, if the smallest value for the dimension D obtained in 95 of 100 (or, better, 950 of 1000) replicates is D_0, then D is greater than D_0 with probability 0.95 under that hypothesis.

6.5 Stationarity

One important characteristic of fractal processes is that their increments are stationary, that is, independent of time. This means that a subsequence of a fractal time series has the same fractal exponents as the original series. Similarly, the increments are independent of the current value of the process. Both conditions are readily tested. For example, a 40-year time series of earthquakes in Japan (1988) is shown to be stationary in Section 8.2. In contrast, time series of bird populations examined in Chapter 11 are not fractal because population increments depend upon the current population value.

Stationarity in fractal patterns is defined similarly and is readily tested visually. Consider, for example, a pattern of islands in a rectangle in the plane. If the pattern is fractal, and the rectangle is divided into four smaller rectangles, then the islands in each of the patterns will have the same general appearance and in particular the same fractal exponents.

Processes which are independent of spatial location should generate stationary patterns. This simple observation can be used to find differences in ecosystem or other dynamics, see the exposition of Hastings *et al.*'s (1982) work on the Okefenokee Swamp in Chapter 10 below. More generally, Meltzer (1990) and Meltzer and Hastings (1992) found evidence for changing grassland dynamics in Zimbabwe in corresponding changes in fractal exponents; see also Chapter 10.

Finally, one must be careful with fractal analysis of short time series just like any other statistical techniques. The above methods for looking for stationarity should prove useful: one need only apply the analytical techniques to parts of the series as well as the whole series.

6.6 Extensions of fractal behaviour

Many natural patterns display fractal behaviour only over a limited range of scales. Mandelbrot (1977, 1982) considered the dimension of a loose ball of string over various scales. In particular, the ball appears three-dimensional at scales large enough to encompass many strands of string, but small compared with the diameter of the ball, one-dimensional at smaller scales which encompass only a single strand, but are large compared with the diameter of a strand, and again three-dimensional at scales small compared with the diameter of a strand. How can we decide between the single scaling region of fractals and such multiscaling behaviour? What other natural generalizations of fractals occur?

We have shown how to fit power laws to experimental data with log transforms and linear regression, and introduced the coefficient of correlation, one measure of the goodness of fit. Unfortunately, the coefficient of correlation can be high in many cases where a linear model (for log-transformed data, and thus a power law model for the original untransformed data) is not appropriate. The definitions of fractal exponents almost invariably (except for the area–perimeter exponent) yield monotone functions, which usually give high correlation.

Fortunately, the existence of multiple scaling regions can frequently be detected by graphing the data, and looking for patterns among the residuals (differences between data points and the regression line)—the residuals will be independent in each scaling region. Such patterns can frequently be detected by statistical tests such as the Durbin–Watson test (Johnston 1973, pp. 249–59; Draper and Lawrence 1981; see Meltzer and Hastings 1992, or Chapter 10, for ecological applications). The Durbin–Watson test measures correlations among residuals u_i with the statistic

$$\sum (u_i - u_{i-1})^2 / \sum (u_i)^2. \qquad (6.25)$$

If the residuals are independent and identically distributed, then $E[(u_i - u_{i-1})^2] = 2E(u_i^2)$ and the Durbin–Watson statistic is approximately 2. The Durbin–Watson statistic will be less than 2 if the residuals display positive serial correlation; see Draper and Smith (1981) for details.

Alternatively, if log-transformed data appear to fit a piecewise linear model, linear regression can be used on separate intervals, overlapping only at endpoints, and the endpoints can be varied so as to minimize the mean square error. If linear regression then yields significantly different slopes on adjacent regions, one should reject the fractal hypothesis of a single power law, and replace it with the *multiscaling* hypothesis of separate power laws over separate regions.

Multifractality represents another possible deviation from fractal behaviour. Consider a fractal pattern of islands formed by undersea mountains, or by flooding a mountain range. More formally, following the construction of Mandelbrot–Weierstrass fractals in Section 5.6, consider the surface

$$z = f(x, y) \tag{6.26}$$

and the pattern of islands defined by

$$\{(x, y) : f(x, y) \geq c\}. \tag{6.27}$$

The boundaries of these islands are the level curves defined by

$$\{(x, y) : f(x, y) = c\}. \tag{6.28}$$

(Topographical maps represent the surface of the earth by similar families of level curves.) If the surface defined by the equation (6.26) is a fractal, then the fractal dimensions of the islands are independent of the sea level c. Conversely, if the above construction yields fractal islands whose fractal dimension depends upon c, then the surface is said to be *multifractal*. Lovejoy and Schertzer (1991) found that strong rainfalls clustered more closely than weak rainfalls; more formally, the dimension of the set of times when the rain fell at an intensity of at least c decreased as c was increased. Thus these patterns are multifractal. In contrast, the scaling behaviour of the Bak–Tang–Weisenfeld (1987) earthquake model and of some real earthquakes appears independent of an intensity threshold (see Hastings and Troyan 1991), and thus is not multifractal. It may, however, be multiscaling—further analysis is needed.

Spatial anisotropy represents yet another deviation from fractal behaviour. The dimension of a section of a fractal surface is independent of the choice of section—thus fractals are isotropic in space. The existence of spatial anisotropy can lead to more complex models involving an elliptic dimension (cf. Lovejoy and Schertzer 1991), or simply lead to a search for deviations from fractal behaviour. Spatial anisotropy can be checked by computing the fractal exponents associated with sections through spatial patterns: for spatially anisotropic patterns, the fractal exponent is independent of the direction of the section.

Our work on earthquake time series (Section 8.2) includes additional tests for stationarity of time series, of which the simplest is to compute the fractal exponents separately for segments of the time series. The exponents of each segment are the same for a stationary time series. Hastings *et al.* (1982) tested for stationarity in spatial patterns similarly, by computing the Korcak exponent for spatially distinct subsets of vegetation 'islands': the fractal dimension of part of a spatially stationary pattern is independent of the area selected.

In conclusion, although many authors have simply assumed that fractal models apply to the system under study, we feel that it is important to demonstrate their applicability. Applicability can be demonstrated by statistical methods, and also by a qualitative search for the scale-invariance of fractals. The absence of natural scales (at least over limited regions) in physical models implies scale-invariance (over these regions), and their presence can prevent scale-invariance.

7

Alternative models

7.1 Introduction

We briefly describe several alternative models for dynamics and time series and compare them with fractal models: cellular automaton models (see Wolfram 1984, 1985, 1986), linear models, including stochastic linear models; low-dimensional nonlinear prediction (Sugihara and May 1990b) based on low-dimensional chaos (see Grassberger and Procaccia 1983). Cellular automaton models are used in Section 8.2 (on earthquake models) and Chapter 10 (on vegetative ecosystems). Chapter 11, on models for single-species population dynamics, compares the use of fractal models, low-dimensional nonlinear prediction, and linear models for predicting local species extinction.

7.2 Underlying dynamics

The purpose of this section is to briefly explore a class of dynamical models which can generate fractal patterns, namely cellular automaton, or CA, models (Wolfram 1984, 1985, 1986). CA models are discrete models for evolution in space and time. Time is usually represented by the nonnegative integers. The cells in a one-dimensional CA model space are usually represented by a finite set $\{0, 1, ..., n - 1, n\}$ of consecutive integers. In a two-dimensional CA model, space is a chessboard corresponding to a set of points with integral coordinates. At any time t, each cell is assigned one state from a (usually finite) set of possible states. The state of each cell at time $t + 1$ depends upon the states of one or more cells at time t. Levin and Paine (1974) developed a CA model, which they called patch dynamics, in order to model the dynamics of immigration and extinction in space and time. We shall develop this model further in Chapter 9.

Even one-dimensional CA models can exhibit considerable dynamical complexity (Wolfram, 1984), including point and periodic attractors, fractals, and chaos. Here is an example, a one-dimensional analogue of Conway's (see Gardner 1983) game of life. This CA has two states: 0 (vacant) and 1 (occupied). The system evolves according to the next-state formula

$$u(x, t+1) = \begin{cases} 0 & \text{if } u(x-1, t) + u(x, t) + u(x, t+1) = 0 \text{ or } 3, \\ 1 & \text{if } u(x-1, t) + u(x, t) + u(x, t+1) = 1 \text{ or } 2. \end{cases}$$

This rule is a toy example of density-dependent growth: a given cell will be occupied 'next year' if some but not all of its neighbours are occupied 'this year'. The one-dimensional game of life is thus what Wolfram (1984) calls a 'totalistic' CA: the state of any cell at time $t+1$ depends only upon the number of occupied cells in a given neighbourhood of it at time t.

The one-dimensional game of life evolves into a Sierpiński triangle (a fractal with dimension log 3/log 2; see Mandelbrot 1977, 1982) from a single source of two neighbouring occupied cells; random initial conditions can yield very complex patterns; see Fig. 7.1.

The original two-dimensional game of life is more complex and not totalistic: a cell will be occupied at time $t+1$ under either of the following conditions: (i) the cell is occupied at time t, and 2 or 3 of the other 8 cells in the 3×3 square centred on it are also occupied; or (ii) the cell is vacant at time t, but 3 of the other 8 cells in the 3×3 square centred on it are occupied.

The Bak–Tang–Weisenfeld (1987) sandbox model for earthquakes, described in Chapter 8 below, and many neural network models (see Linsker 1988) are also CA models.

REMARKS 7.1 There is a close relationship between some CA models and related partial differential equation (PDE) models. The numerical solution of PDE's on finite grids yield cellular automata with real-valued states. Consider, for example, the basic diffusion equation in one space dimension

$$\frac{\partial u}{\partial t} = \frac{1}{2} \frac{\partial^2 u}{\partial x^2}. \tag{7.1}$$

We may discretize (7.1) by replacing $\partial u/\partial t$ with the difference quotient $[u(x, t+h) - u(x, t)]/h$ and replacing $\partial^2 u/\partial x^2$ with the similar difference quotient $[u(x+k, t) + u(x-k, t) - 2u(x, t)]/k$, where h is the time step and k is the space step. If we set both the time and space steps equal to 1, we obtain the discrete equation

$$u(x, t+1) - u(x, t) = \tfrac{1}{2}[u(x+1, t) + u(x-1, t) - 2u(x, t)],$$

or, more simply,

$$u(x, t+1) = \tfrac{1}{2}[u(x+1, t) + u(x-1, t)]. \tag{7.2}$$

Conversely, the diffusion limits of suitable CA models yield partial differential equations. The mathematics is analogous to the usual derivation of the

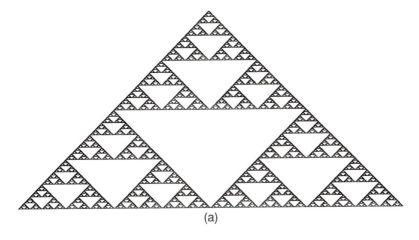
(a)

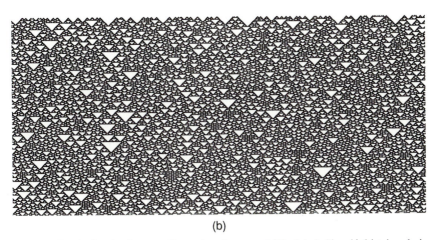
(b)

Fig. 7.1 Evolution in the one-dimensional game of life. (a) A Sierpiński triangle is generated starting from two adjacent occupied cells. (b) Random initial conditions yield complex patterns.

heat equation (see Kreysig 1988). Levin and Paine (1974) took this approach to modelling ecosystem dynamics; see also Chapter 10 below.

7.3 Linear models

The mathematical study of dynamics was first motivated by the classical physics of small systems. This study led to the parallel development of classical mathematics (calculus and differential equations) and classical physics during the seventeenth to nineteenth centuries. The mathematical models included both systems of differential equations

$$\frac{dx}{dt} = g(x, t) \tag{7.3}$$

and their analogues in discrete time

$$x(t + h) = x(t) + hg(x, t) + \cdots = f(x, t), \tag{7.4}$$

where $x(t)$ represents the *state* of the system at time t. We shall consider the discrete-time system (7.4), and assume, without loss of generality, that the time step $h = 1$. A system (7.4) is called *autonomous* if its evolution does not depend explicitly upon time, in which case we have

$$x(t + 1) = f(x(t)). \tag{7.5}$$

A state x_{eq} is called an *equilibrium* if the system remains in the state x_{eq} whenever it is started there:

$$x_{eq} = f(x_{eq}). \tag{7.6}$$

The equilibrium x_{eq} is called stable if the state approaches x_{eq} whenever the system is started sufficiently near x_{eq}. That is, there is an $\varepsilon > 0$ such that $\|x(0) - x_{eq}\| < \varepsilon$ implies

$$\lim_{t \to \infty} x(t) = \lim_{t \to \infty} f(f(\ldots f(x(0)) \ldots)) \quad (t \text{ iterates of } f)$$

$$= x_{eq}. \tag{7.7}$$

For most classical systems, the *next-state* function f can be expanded about the equilibrium x_{eq} in a convergent Taylor series

$$f(x) = f(x_{eq}) + [\partial f_i/\partial x_j](x - x_{eq}) + (\text{higher-order terms}), \tag{7.8}$$

where $[\partial f_i/\partial x_j]$ is the *Jacobian* matrix of partial derivatives of f. We note that

$$f(x_{eq}) = x_{eq}, \tag{7.9}$$

and rewrite equation (7.8) in the following form:

$$f(x) - x_{eq} = [\partial f_i/\partial x_j](x - x_{eq}) + (\text{higher-order terms}). \tag{7.10}$$

If the size of the dominant eigenvalue (largest eigenvalue in size) of the matrix $[\partial f_i/\partial x_j]$ is *not* equal to 1, then the dynamics of the next-state function can be approximated in a sufficiently small neighbourhood of x_{eq} by neglecting the 'higher-order terms' in equation (7.10) and writing

$$f(x) - x_{eq} = [\partial f_i/\partial x_j](x - x_{eq}). \tag{7.11}$$

This process is called *linearization* and the corresponding *linear model* (7.11) may be represented more simply by letting

$$\boldsymbol{u} = \boldsymbol{x} - \boldsymbol{x}_{\text{eq}} \quad \text{and} \quad M = [\partial f_i / \partial x_j] \qquad (7.12)$$

and writing

$$\boldsymbol{u}(t + 1) = \boldsymbol{g}(\boldsymbol{u}(t)) = M\boldsymbol{u}(t). \qquad (7.13)$$

Even though linear models involve significant simplifications of the dynamics, they frequently give interesting insights into the behaviour of complex systems such as ecosystems (May 1974). For example, population levels in many ecosystems appear relatively stable, and much of the theory of eigenvalues of random matrices was developed in order to understand the (Lyapunov) stability of these systems (May 1972, 1974). See also Hastings (1983).

Models such as (7.13) are *deterministic* in that they involve no random components. Hastings (1983) extended the models of May (1974) by introducing random walk terms $\Delta w(t)$, yielding the linear *stochastic* model

$$\boldsymbol{u}(t + 1) = M\boldsymbol{u}(t) + \Delta w(t), \qquad (7.14)$$

and studying the relationship between the eigenvalues of M and the asymptotic statistics of $\boldsymbol{u}(t)$. See also Chapter 11.

REMARKS 7.2 In the special case that M is the identity matrix, equation (7.14) is just a random walk. If M has spectral radius $\rho(M)$ less than 1, then equation (7.14) has a stable equilibrium at 0, and at large times $\boldsymbol{u}$ is normal with expected value 0 and variance

$$E((\Delta w)^2)/\{1 - [\rho(M)]^2\}.$$

Many systems without stable equilibria do display stable behaviour. For example, many systems including certain Lokta–Volterra models for predator–prey interactions (see May 1974) yield stable cyclic behaviour. Roughly, a cycle is stable if the trajectories of nearby points approach and track the cycle as the time approaches ∞ (see Hirsch and Smale 1974). Some Lokta–Volterra models for predator–prey interactions display stable cycles (see May 1974) although the cycles in the original model were only neutrally stable (trajectories of nearby points remained close to the cycle but did not approach it).

In 1963, E. N. Lorenz described complex stable *aperiodic* behaviour in a model for fluid flow—this well-studied example has been termed the Lorenz attractor. Stable equilibria and stable limit cycles are classical examples of *attractors*. A compact subset A of the state space of a dynamical system is called an attractor provided that: (i) A is invariant (if the state of the system starts in A, then it remains in A); (ii) trajectories which start sufficiently near

A approach *A*; and (iii) *A* is minimal (no proper closed subset of *A* has these properties). See Hirsch and Smale (1974) for mathematical details. The study of attractors is important because the geometry of an attractor frequently captures much of the dynamics on that attractor—for example, in continuous-time models, stable cycles correspond to attractors which look like (more formally, are homeomorphic to) circles, and, in particular, are *one-dimensional*. We shall describe techniques for finding and studying low-dimensional attractors in the next section.

7.4 Low-dimensional nonlinear models

Many natural systems and nonlinear mathematical systems display more complex stable behaviour, as shown by the existence of attractors other than equilibria and limit cycles. E. N. Lorenz (1963) found perhaps the first such behaviour in a mathematical model of a natural system—a *strange attractor* in a truncated version of the Navier–Stokes equations for flow in the atmosphere. Linear models cannot capture the dynamics of strange attractors. Thus an alternative approach is needed to tame the nonlinearities.

The discrete-time logistic equation (see May 1974) of population dynamics provides an instructive example. In this model

$$y(t) = ry(t-1)[1 - y(t-1)],\qquad(7.15)$$

where the state variable $y(t)$ represents the population level at time t in suitable arbitrary units, the parameter r represents the intrinsic growth rate, and the nonlinearities are due to the density-dependent factor $1 - y(t-1)$. (Equation (7.15) is usually written in the form $y(t+1) = ry(t)[1 - y(t)]$; our unusual notation is chosen for Fig. 7.2 below.) For simplicity, we shall rewrite equation (7.15) in the form

$$y_{\text{new}} = ry(1-y).\qquad(7.16)$$

This model shows a wide variety of simple and complex behaviours depending upon the value of the intrinsic growth rate r (May 1974). We restrict r to the interval $0 \leqslant r \leqslant 4$ in order that equation (7.16) define a mapping from the closed unit interval $0 \leqslant y \leqslant 1$ to itself. For $r < 3$, there are stable equilibria given by the formula

$$y_{\text{eq}} = \max\{0, 1/(r-1)\}.\qquad(7.17)$$

For $r > 3$ the situation is significantly more complex. As r is slowly increased, first the point attractor y_{eq} becomes unstable and bifurcates into an attractor consisting of a pair of points of period 2. Later one obtains similar periodic attractors of period 4, 8, 16, etc. Finally, for $r > 3.57$ there are many regions which appear to exhibit complex behaviour (May 1974; Li and Yorke 1975),

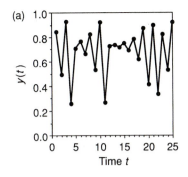
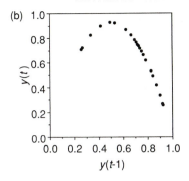

Fig. 7.2 Two views of the discrete-time logistic equation. (a) A graph of the time series of 25 points generated by the equation $y(t) = ry(t-1)$ shows no apparent pattern. (b) Graphing the corresponding 24 points $(y(t-1), y(t))$ clearly shows the original function $y(t) = ry(t-1)$.

including chaos, a stable form of apparently random behaviour. However, other parameter values, such as $r = 3.83$, give stable periodic points. For example, a stable period-3 cycle is born at $r = 3.83$, bifurcates to give period 6, 12, etc. as r is increased, and completely disappears by the time $r = 3.85$ (Smale and Williams 1976). See also Collet and Eckmann (1980) for a fuller description of the dynamics of the discrete-time logistic equation.

On the surface, for many values of $r > 3.57$, the time series $\{y(t)\}$ of population levels (values of y) generated by iterating equation (7.16) appears indistinguishable from a random sequence. However, if we search for nonlinear correlations among the population values by plotting ordered pairs of successive values $(y(t-1), y(t))$, then the deterministic nature of the discrete-time logistic model becomes evident.

Grassberger and Procaccia (1983) generalized the ideas behind Fig. 7.2 with their algorithm for computing the correlation dimension. Suppose that one is given a time series $\{x(t)\}$. For each positive integer n, plot all n-tuples of consecutive values

$$(x(t), x(t+1), \dots, x(t+n-1))$$

in n-dimensional *ambient* Euclidean space, and determine the *embedding* dimension D (cluster dimension or Hausdorff dimension) of the resulting pattern. If D is less than n, then the n-tuples $(x(t), x(t+1), \dots, x(t+n-1))$ do not correspond to randomly selected points in n-dimensional Euclidean space. Moreover, once n is sufficiently large that D stabilizes, the embeddings in n-dimensional and $(n+1)$-dimensional Euclidean space have the same dimension D, and this implies a nonlinear dependence of $x(t+n)$ upon the n-tuple

$$(x(t), x(t+1), \dots, x(t+n-1)).$$

Table 7.1   Summary of three approaches for modelling time series

Model:	Linear (plus noise)	Low-dimensional nonlinear	Fractal
Basic approach:			
	linear stochastic	nonlinear deterministic	random descriptive
Predictability of next event:			
	yes	yes	no
Long-term predictability:			
	yes	no	no
Scale-invariant:			
	no	sometimes	yes
Scale-invariant over scaling regions:			
	sometimes	sometimes	yes
Grassberger–Procaccia dimension of time series (after detrending):			
	high	low	high
Amount of data needed to fit:			
	little	large	relatively little

This dimension is essentially the number of degrees of freedom in a nonlinear sense. Moreover, this construction has essentially found a set of coordinates for parametrizing the attractor: to each point in the attractor we associate the n-tuple $(x(0), x(1), \ldots, x(n-1))$ resulting from n successive observations of the quantity x.

Sugihara and May (1990b) used these ideas to develop a method for distinguishing low-dimensional chaos from measurement error in natural time series. The method has been fruitfully applied in many areas, ranging from modelling disease outbreaks to long-range weather forecasting.

REMARKS 7.3 In theory, each value generated by a pseudo-random number generator should be statistically independent of all previous values, and thus the set of n-tuples generated by a pseudo-random number generator should have dimension n. In essence, Marsaglia (1968) used this fact as a test of pseudo-random number generators. See also Knuth (1981). However, this behaviour is difficult to achieve since pseudo-random number generators are deterministic algorithms.

Caution. The Grassberger–Procaccia algorithm requires a relatively long time series (of the order of 10^D where D is the dimension of the underlying attractor) points (Ghil *et al.* 1991; Ruelle 1990; Smith 1988). However, many natural time series are high-dimensional or short or both. In this case one is forced into the world of statistical models, including random fractals.

Three approaches for modelling time series are shown in Table 7.1.

8

Examples

8.1 Introduction

In this chapter we introduce the application of fractals to the study of spatial patterns through several illustrative examples: the work of Hastings and Troyan (1991) on scaling in earthquake time series (Section 8.2), the work of Hastings *et al.* (1992) on pancreatic islets (Section 8.3), and the work of Caserta *et al.* (1990) on neuronal processes (Section 8.4). The following section considers the possibility or impossibility of predicting earthquakes, and asks about possible measurable differences between earthquakes along fault lines and 'mid-continent' earthquakes far from fault lines. Which comes first—the fault or the earthquake? The last two sections investigate a possible universal role for diffusion-limited aggregation in developmental biology (Witten and Sander 1981; Meakin and Tolman 1989), as suggested by Caserta *et al.* (1990), Kleinfeld *et al.* (1990), and Hastings *et al.* (1992).

8.2 Earthquake models

We investigate scaling properties of some real and artificial earthquake time series, such as the series of Japanese earthquakes shown in Fig. 8.1. This investigation is motivated by the Bak–Tang–Weisenfeld (1987) 'sandbox' model, a CA model thought to capture the main phenomenological aspects of earthquake dynamics. The model formalizes the following thought experiment. Consider a sandbox in which sand is added slowly and randomly. Where the slope of sand (representing the 'local strain') exceeds a given critical value, sand falls from high points to neighbouring low points, thus reducing the local slope. Cascades arise whenever this readjustment causes the local slope to exceed the critical value at new points. The model thus captures the dynamics of energy storage as strain, energy release when the strain exceeds a critical value, and cascades as strain release at one point overstresses the system at nearby points.

The sandbox model is readily formulated as a two-dimensional cellular automaton. Cells are located at lattice points in a region in the plane. The state $z(x, y)$ of each cell (x, y) takes on the values 0, 1, 2, 3, and corresponds to the *strain* at the point (x, y). Sand is added to randomly and independently chosen cells, and the corresponding increase in strain is represented by the

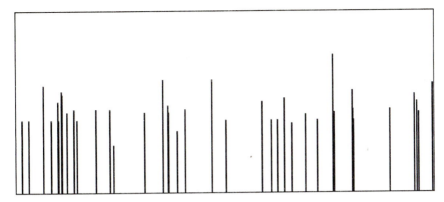

Fig. 8.1 Major earthquakes in Japan. A graphical representation of the dates (on the horizontal axis) and magnitudes (ranging from 4 to 9 on the vertical axis) of events in the first 2400 days in the catalogue in Utsu (1988). Only damaging events or events with magnitude at least 6.0 are shown. What patterns are present in these data?

state transition $z \to z + 1$. The addition of sand may cause the state of a cell to temporarily exceed the value of 3, in which case sand topples onto nearby cells, representing the local release of strain. This is represented by the state transitions

$$z(x, y) \to z(x, y) - 4,$$
$$z(x', y') \to z(x', y') + 1 \text{ at the four neighbours of } (x, y). \Big\} \qquad (8.1)$$

The Bak–Tang–Weisenfeld model has no natural scale, making its dynamics scale-invariant, except possibly at scales comparable with the size of the 'sandbox'. Therefore the model is scale-invariant, and thus any pair of related descriptive parameters must be related by a power law. For example, the Bak–Tang–Weisenfeld model predicts the Gutenberg–Richter (1942, 1956a, b) law, a power law for the distribution of energy release earthquakes:

$$\text{number(energy release} > E) = \text{const} \times E^{-c}, \qquad (8.2)$$

or, equivalently, since the magnitude M is the logarithm of the energy release:

$$\log[\text{number(magnitude} > M)] = \text{const} - cM \qquad (8.3)$$

How well does the Bak–Tang–Weisenfeld model capture the dynamics of real earthquakes? We approach this quesion by studying the scaling properties of real and simulated earthquake time series: see Tang and Bak (1988) and Erzan and Sinha (1991) for model results. As an application, we also describe the consequences for predictability of earthquakes. In particular, we test real and model data for stationarity (do segments of the series have the same statistics, and in particular the same exponents, as the whole series?), number of scaling regions, and monofractal versus multifractal behaviour (are scaling properties of earthquake time series independent or dependent of a minimum magnitude threshold M_c)?

We also ask whether there are good alternative descriptions (for example, exponential waiting time models) for the time series? The *seismic gap hypothesis* 'states that earthquake hazard increases with time since the last large earthquake on certain faults or plate boundaries' (Kagan and Jackson 1991). The seismic gap hypothesis is consistent with an exponential waiting time model, but not with a fractal model, which implies a relatively constant hazard in each area. Kagan and Jackson (see also the review Monastersky 1992) recently found that earthquakes frequently are more likely to recur near previous sites, apparently negating the seismic gap hypothesis and confirming the fractal hypothesis.

8.2.1 *Methods*

We computed the Korcak exponent B associated with the distribution gaps between earthquakes:

$$\text{number}(\text{gap} > t) = \text{const} \times t^{-B}. \tag{8.4}$$

(Following standard techniques, log-transformed data were fitted to the log-transformed equation $\log[\text{number}(\text{gap} > t)] = \text{const} - B \log t$.)

Stationarity was tested by computing the Korcak exponent for portions of the time series. Multiscaling was tested by comparing data points with regression lines, and by considering exponential waiting time models. Multifractality was tested by restricting the time series to earthquakes of magnitude greater than a critical magnitude and looking for effects upon the Korcak exponent. Both aspects could be tested at the same time—for example, the time series of Utsu (1988), illustrated in Fig. 8.1, consists of 600 earthquakes, of which 536 have magnitude at least 6.0. The series of 536 events was subdivided into shorter series of 101 events (100 gaps each). We also used computer simulations to generate artificial time series with power law gaps and thus determine confidence limits for the measurement of B, and to generate a time series from the sandbox model for further comparison.

8.2.2 *Results*

We begin with an empirical study of the distribution of Korcak exponents, using computer simulation, in order to establish statistics for later comparison. Results are given in Tables 8.1 and 8.2.

Table 8.1 Distribution of the Korcak exponent for simulated series of length 100, based on 25 replicates of each series

Nominal B	Observed B (mean ± s.d.)	Minimum	Maximum
0.8	0.81 ± 0.13	0.57	1.08
1.0	1.00 ± 0.19	0.70	1.49
1.25	1.28 ± 0.26	0.87	1.98

Table 8.2 The effects of the length of a time series upon its distribution of the Korcak exponent given a nominal exponent of 1.2. The study used 100 replicates of each series

Length	Observed B (mean ± s.d.)	Minimum	Maximum
10	1.06 ± 0.50	0.00	2.64
100	1.16 ± 0.23	0.65	1.74

Using this statistical data, we found the time series under study to be stationary, multiscaling, and monofractal (as opposed to multifractal within each interval). However, we could not rule out an alternative exponential waiting time model. Figure 8.2 illustrates the multiscaling and exponential time models, and Table 8.3 the tests for multifractality and stationarity.

Note that large events appear to be a random subset of all events. We also found no special behaviour in neighbourhoods of large events. Moreover, computer simulations (sandbox) and real time series looked visually similar at time scales long compared with the addition of sand.

8.2.3 *Discussion*

Our fractal hypothesis appears consistent with Kagan and Jackson (1991), but more work is needed. In particular, one should investigate time series of 'mid-continent' earthquakes (that is, events far from fault lines), as well as quakes along or near known faults, to see if there are any differences.

What are the implications for predictability if the data are fractal? The hyperbolic distribution of gaps (between events) is a very 'broad' distribution, and thus it is hard to determine the Korcak exponent from short time series of empirical data. This implies sharp limits on both the predictability

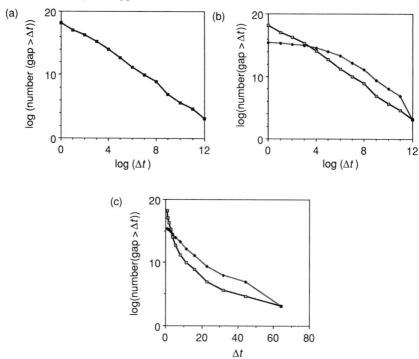

Fig. 8.2   Models for earthquake time series. (a) A log–log plot of the number of gaps between earthquakes of magnitude at least 6.0 as a function of their duration, showing a good fit to a Korcak-type power law. (b) A similar log–log plot for earthquakes of magnitude at least 6.5 (dashed line) shows significant deviations from such a power law. The data for magnitude at least 6.0 are shown as a solid line. (c) A similar plot of the log of the number of gaps as a function of their duration, showing a possible exponential waiting time distribution.

of single large events and the determination of their statistics from time series alone. In particular, since large events are rare by the Gutenberg–Richter law, it is especially difficult to determine whether there is an excess number of large events, a highly debated question (Gutenberg and Richter 1942, 1954, 1956a, b; Carlson and Langer 1989). It would be interesting to apply the above analysis to data from other regions, and in particular to compare events at fault lines with events in the interiors of continental plates.

The sandbox model offers a reasonable fit to observed data. It would be interesting to study the dynamics of an anisotropic sandbox model (for example, topple at $z \geqslant 6$, moving 2 grains north, 2 south, 1 east, 1 west) which incorporates fault zones.

Table 8.3 Tests for multifractality and stationarity: the Korcak exponent for the time series of Japanese earthquakes, and for subseries defined by considering only those earthquakes with a given minimum magnitude, and portions of the above series

Minimum magnitude	Events numbered	Korcak exponent B
0	1 to 599 (all)	1.22
	1 to 100	1.11
	101 to 200	1.10
	201 to 300	1.31
	301 to 400	1.21
	401 to 500	1.21
	501 to 599	0.67*
6	1 to 535 (all)	1.05
	1 to 100	1.22
	101 to 200	1.06
	201 to 300	1.08
	301 to 400	1.32
	401 to 500	0.96*
	436 to 535	0.65*
6.5	1 to 263 (all)	1.13
	1 to 100	1.19
	101 to 200	1.17
	164 to 263	1.15
7	1 to 114 (all)	1.14

* Statistically significantly different from the Korcak exponent for all the events at or above the listed minimum magnitude, at least at the 0.05 level, using the computer simulation results of Table 8.2.

8.3 Geometry of pancreatic islets

8.3.1 *Background*

We report on results of Hastings *et al.* (1992). Pancreatic islets (containing, among other cell types, insulin-producing β-cells) occupy less than 1% of the pancreatic volume and appear irregularly distributed, even on small scales. It is therefore difficult to understand their underlying geometry from standard methods of analysis of immunohistochemically stained two-dimensional sections (Logothetopolis 1972; Weibel 1979; Hellestrom and Swenne 1985).

8.3.2 Biological materials and methods

Alloxan is a selective β-cell toxin commonly used to generate experimental models of diabetes mellitus in laboratory animals. Unlike other animals, in the guinea pig, the lost β-cell volume is predictably regenerated within five days after the alloxan injection (Gorray *et al.* 1986a, b). The processes of regeneration and islet formation were studied by comparing the observed fractal properties of islet centres in microscopic sections. Preparation and data logging are detailed in Hastings *et al.* (1992).

8.3.3 Are islet patterns fractal?

We first computed the cluster dimension (Section 3.3) of islet centres. The cluster dimension D is the exponent in the power law for the number $n(r)$ of islets within a distance r of a typical islet

$$n(r) = mr^D. \tag{8.5}$$

As usual, we fit the linear model

$$\log n(r) = \log m + D \log R \tag{8.6}$$

to log-transformed data. The parameter m is a 'fractal density' with units $(\text{length})^{-D}$; the typical interislet distance $r = m^{-1/D}$ makes $n(r) = 1$. It is important to check the goodness of fit of the linear model (8.6) using both the coefficient of correlation and some test (perhaps just a simple visual test) for the absence of apparent patterns among residuals (differences between data points and the regression line). In a fractal pattern one expects to see a linear relationship between $\log r$ and $\log n(r)$ as long as r is sufficiently small so that circles of radius r around most islets stay within the confines of the pattern.

  We found a good fit between the data and the linear model (8.6), indicating scale-invariance or fractal behaviour. Islets in 35 sections from control animals had a cluster dimension of 1.56 ± 0.04 (mean $\pm$ standard error of measurement); islets in 4 slides from experimental animals had $D = 1.48 \pm 0.13$ (not significant) (Hastings *et al.* 1992).

  One can now ask whether the observed cluster dimensions are significantly different from 2. In order to answer this question, we also empirically determined D for two types of computer generated patterns: randomly distributed centres in the plane and randomly distributed centres in a typical lower-dimensional fractal (an attractor based on the Sierpiński triangle) with expected $D = \ln 3/\ln 2 = 1.58$. This attractor was constructed with iterated function systems (Barnsley *et al.* 1986; Barnsley 1988); see the program *fractgame* in Section 12.8. Random subsets of the plane and fractal had

$D = 1.91 \pm 0.03$ and 1.74 ± 0.03 (100 replicates of each), respectively. These results suggest that all of the variability in experimental data can be explained by sampling, and thus there is no necessary biological explanation. On the other hand, the combined set of all 39 sections had $D = 1.55 \pm 0.04$, significantly different from the value for random subsets ($p < 0.001$, t-test).

The intersection formula in Section 3.7.3 implies that a planar section of a D-dimensional pattern in three-dimensional Euclidean space has dimension $D - 1$. Thus islet centres in both islet-regenerated experimental animals and controls cluster in similar 2.6-dimensional fractal subsets of the pancreas (see Hastings *et al.* 1992). It is believed that islets form along tips of a tree-like structure of exocrine ductules (Falkmer 1985). Since the dimension of a random (Poisson) subset of a D-dimensional fractal has cluster dimension D, our results suggest the testable hypothesis that exocrine ductules may be a random fractal of dimension 2.6 (Hastings *et al.* 1992).

The observed clustering of islet centres has important consequences for sampling and allometry (Hastings *et al.* 1992). The number of islets in a square of side s scales as $s^{1.6}$, and thus a 2 mm square section will contain on average only $2^{1.6} = 3.03$ (*not* 4) times as many islets as a 1 mm square section. See Weibel (1979, pp. 153–7) and Paumgartner *et al.* (1981) for similar applications to scale-dependent measurements in stereology.

8.4 Dimension of neuronal processes

Caserta *et al.* (1990) and Kleinfeld *et al.* (1990) observed a fascinating fractal structure in developmental biology: the cluster dimension of neuronal processes is about 1.7 ± 0.1.

We thus observed that the processes of nearby neurons can be expected to intersect in a subset of dimension $1.7 + 1.7 - 3 = 0.4$ (see Section 3.7.3) and thus nearby neurons almost certainly intersect. We thus expect micro-columns, small assemblages of neurons in the brain (see Schmidt 1978), to be almost completely connected.

REMARKS 8.1 Our ideas are based on those of Lovejoy *et al.* (1986). They used similar methods to show that the global weather detection network, of fractal dimension 1.75 on the two-dimensional surface of the arth may miss highly clustered intense storms of fractal dimension less than $2 - 1.75 = 0.25$.

The above results suggest several questions, which are beyond the scope of the present work. In particular, how can one explain the observed fractal dimensions? Are they a consequence only of universal properties of growth processes such as diffusion-limited aggregation (Witten and Sander 1981; Meakin and Tolman 1989), as suggested by Caserta *et al.* (1990), Kleinfeld *et al.* (1990), and Hastings *et al.* (1992)? Diffusion-limited aggregation predicts $D = 1.7$ (the value obtained for growth of neuronal processes) in the plane and $D = 2.4$ (pancreatic ductules had $D = 2.5$) in space. Perhaps random growth toward slowly diffusing factors follows such models.

On the other hand, perhaps the dimensions and structures themselves are also dictated by biological functions, such as maximizing oxygen uptake in the bronchi and short-range communication in the brain. In either case, diffusion-limited aggregation may play a central role in pattern formation in developmental biology. Finally, since universal processes such as diffusion-limited aggregation are independent of small changes in parameters, they yield robust patterns which are relatively constant during growth. Such universality may even be favoured in evolution, in order to ensure structure and thus function in the face of noise.

9

Fractal analysis of time series

9.1 Introduction

The purpose of this chapter is to introduce and illustrate the fractal analysis of time series. The central limit theorem in statistics implies a power law for the sum of n (identical, independently distributed, bounded) random variables of mean 0: the size the sum scales as $n^{1/2}$. The displacement in a random walk consisting of n steps scales similarly as $n^{1/2}$. Hurst (1951) asked whether fluctuations in the cumulative discharge of the Nile River, the sum of many incremental discharges, scales similarly and found instead that the range of fluctuations over a time period T scaled as T^H, where H is strictly greater than $\frac{1}{2}$. Mandelbrot (1965), Mandelbrot and Van Ness (1968), and Mandelbrot and Wallis (1969) explain this behaviour in terms of scale-invariant long-term correlations, and thus introduced fractal modelling of time series. We shall develop such fractal models using our own analysis of ocean surface temperatures at the Scripps pier (La Jolla, California, USA) and New Jersey (USA) rainfall data.

We begin by reviewing the process of 'detrending', using to eliminate large cyclic trends (such as those in temperature data) or linear trends, and considering when to form cumulative time series. We next illustrate these processes and fractal modelling by studying California ocean surface temperatures and New Jersey rainfall data. The chapter concludes with a brief formal comparison of fractals with residence time models. A detailed case study on the application of fractals to population dynamics and the possible application to forecasting species extinction (see Sugihara and May 1990a) is given in Chapter 11.

9.2 Detrending

Many time series show pronounced cyclic trends. For example, daily temperature data follow an annual cycle whose magnitude overwhelms other fluctuations. Rainfall data in may areas undergo a similar annual cycle of similar magnitude. It is therefore desirable to compare the temperature on a given date or the amount of rainfall in a given month to the average temperature on that date or the average amount of rainfall in that month. This can be accomplished by detrending.

In both the cases cited above, the dominant cycle has a duration of one year, and it is this cycle which we choose to remove. We assume that there is sufficient data to reliable compute daily or monthly averages, as appropriate. (In the case where the graph of the average temperature on each day seems to fluctuate excessively rather than follow a smooth curve, one can readily apply further smoothing techniques such as computing the moving average (over the date variable) of the average temperature by days.)

The deviation from this trend can be measured as the difference between actual values and the tend, or as the ratio of actual values to the trend. In the case of temperature data, subtracting the corresponding trend value from each data point seems clearly most appropriate. The situation for rainfall data is less clear, but questions about flood control and water storage concern total amounts of water and not relative amounts of water. We therefore chose to subtract out the trend data. (There is one caveat in doing the subtraction. In a rainy month, for example a month with 6 inches of rainfall, an excess of 1 inch may not seem significant. However in a dry month, for example a month with 1 inch of rainfall, an excess of 1 inch represents a doubling of the amount of rain.)

Although actual (noncumulative) deviations are bounded, and thus cannot be fractal, the cumulative deviation may be fractal, just as a random walk with bounded increments is fractal. (The increments of a fractal process taken over a time step Δt are Gaussian and thus 'effectively bounded'.) We therefore studied cumulative deviations, just as in Hurst's (1951, 1956) study of the *cumulative* river discharges of the Nile river. There is another useful test for fractal behaviour involving the range of the Hurst exponents, $0 < H < 1$, and the equivalent restriction of the coefficient of correlation ρ between successive increments to the interval $-\frac{1}{2} < f < 1$ (see also Section 4.3). If experimental results should lie outside these ranges, then one should consider taking sums or differences.

Since annual river discharges, daily temperatures, and instantaneous heart rates are bounded, they cannot be fractal processes themselves. However, they could be the increments of fractal processes over time steps of 1 year, 1 day, and 1 sampling interval, respectively.

9.3 La Jolla, California, temperature data

We analysed a time series of over 20 000 daily readings of ocean surface temperature from the end of the Scripps pier in La Jolla, California, from the years 1927 to 1989. Parts of this data have been previously analysed for spectral properties (Tont 1975, 1981).

9.3.1 *Methodology*

We began by looking for annual trends by computing daily average temperatures over the more than 60 years of data. Leap years were handled by deleting every 1461th temperature ($4 \times 365.25 = 1461$). We then further smoothed the daily average temperature by applying a seven-day moving average. This yielded average temperatures over 365-day 'years'. We found an annual cycle and removed it by subtracting the average temperature at each date from the corresponding actual temperature.

We looked for fractal behaviour by local growth of variance method, that is, by computing the correlations between successive increments of the cumulative temperature difference function $F(t)$ defined by

$$
\left.
\begin{aligned}
&F(0) = 0, \\
&F(t + 1) = F(t) + \text{(temperature on day } t + 1) \\
&\qquad - \text{(average temperature on day } t + 1).
\end{aligned}
\right\}
\tag{9.1}
$$

We then computed the correlation between successive increments $F(t + h) - F(t)$ and $F(t + 2h) - F(t + h)$ for lags h of from 1 day to 2500 days. Since the temperature data was detrended, we could assume that the mean difference between excess temperature was 0, and compute the coefficient of correlation by the methods of Section 4.4.

REMARKS 9.1

(a) The increment $F(t + h) - F(t)$ is the sum of the temperature deviations on days $t + 1, t + 2, \ldots, t + h$, and may be considered as the 'cumulative temperature deviations'. In fact, $F(t + h) - F(t)$ is the average excess temperature over the h days above, and the correlations are correlations between excess temperatures in successive periods of h days.

(b) The cumulative temperature deviation $F(t)$ can grow large in size, in contrast with the temperature itself, which is surely bounded, for example between 0 and 50°C!

(c) We also tried computing correlations between differences between excess temperatures, for example [(temperature at day $t + h$) − (average temperature at day $t + h$)] − [(temperature at day t) − (average temperature at day t)], but obtained correlations which rapidly tended towards $-\frac{1}{2}$.

9.3.2 *Results*

Increments in the cumulative temperature deviation are strongly positively correlated (as might be expected) for very short lags of one to several days. This correlation is gradually lost to a near-zero correlation at roughly two years. The correlation continues to decrease slowly with increasing time lag,

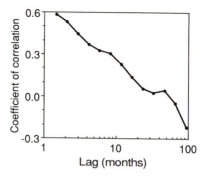

Fig. 9.1   Correlation of successive increments in a time series of almost 80 years of ocean surface temperatures from the Scripps Institute of Oceanography pier, La Jolla, California, USA.

and becomes significantly negative at times of six or more years. In terms of the fractal exponent H, we found that H is greater than $\frac{1}{2}$ for short lags, decreases to about $\frac{1}{2}$ for times from two to five years, and becomes less than $\frac{1}{2}$ for periods of six years or more. The value $H = \frac{1}{2}$ is the Brownian value; we interpret $H > \frac{1}{2}$ as 'persistence' and $H < \frac{1}{2}$ as 'anti-persistence' or a tendency towards reversal. These results mean that if one month is warmer than average then the next month is quite likely to be warmer than average. If one year is warmer than averge, the next year is somewhat likely to be warmer than average. This tendency is lost after about two years, and reverses after about six years (see Fig. 9.1).

REMARKS 9.2 This behaviour may be related to the duration of El Niño events (Rasmussen *et al.* 1990; Ghil *et al.* 1991) which characteristically last for about two years, and might thus be explained by residence time models of Section 9.5. A system with a residence time of the order of two years would smooth out anti-persistent behaviour on shorter time scales, but have little effect on longer time scales.

A toy Fourier analysis might provide additional hints. Consider a system with a broad spectral peak, and calculate the correlation ρ between successive increments as a function of the lag Δt—the correlation method of Section 4.3. Without loss of generality, we use units in which the spectral peak has time scale 2π. Then the correlation ρ will be 0 when

$$\int_0^{2\pi} [\sin(t + 2\Delta t) - \sin(t + \Delta t)][\sin(t + \Delta t) - \sin t] \, dt = 0. \quad (9.2)$$

Evaluating the above integral yields

$$\int_0^{2\pi} [\sin(t + 2\Delta t) \sin(t + \Delta t) - \sin^2(t + \Delta t) - \sin(t + 2\Delta t) \sin t \\ - \sin(t + \Delta t) \sin t] \, dt$$

$$= \int_0^{2\pi} [-\sin 2(t + \Delta t) - \sin(t + 2\Delta t) \sin t] \, dt$$

(since $\int_0^{2\pi} \sin(t + 2\Delta t) \sin(t + \Delta t) \, dt = \int_0^{2\pi} \sin(t + \Delta t) \sin(t) \, dt$)

$$= -\pi - \int_0^{2\pi} \sin(t + 2\Delta t) \sin t \, dt$$

$$= -\pi - \int_0^{2\pi} [\sin^2 t \cos 2\Delta t + \sin t \cos t \sin \Delta t] \, dt$$

$$= -\pi - (\cos 2\Delta t) \int_0^{2\pi} \sin^2 t \cos 2\Delta t \, dt$$

(the second integrand above integrates to 0)

$$= -\pi - (\cos 2\Delta t)\pi. \tag{9.3}$$

Thus the correlation is 0 when $\cos 2\Delta t = -1$, or when $\Delta t = \pi/2$, one-quarter of the time scale of the natural peak. The graph of the sine function is also relatively large (over 70% of its maximum) over one-quarter of its cycle, and thus the time scale of zero correlation is approximately the time scale of maxima of the sine function. Since El Niño events may occur on a strange attractor (Rasmussen *et al.* 1990; Ghil *et al.* 1991), one expects more complex quasi-periodic behaviour.

9.4 New Jersey rainfall data

We performed a similar analysis on 20 years of almanac data on New Jersey rainfall after discussions of flood control projects of the Army Corps of Engineers in the area.

9.4.1 *Methodology*

We analysed these data as follows. We first computed monthly average rainfalls over the 20 years of data. There was no pronounced annual trend, so we did not detrend the data. We then used the ρ–lag method as above, for lags of one to 24 months. The maximum lag was limited so that there would be sufficient data to reliably compute the *population* coefficient of correlation from experimental data.

9.4.2 *Results*

As above, the rainfall data display short-term persistence, middle-term Brownian behaviour, and long-term anti-persistence. The short-term persistence (positive correlations and thus $H > \frac{1}{2}$) holds for periods of up to five months. Brownian or approximately Brownian behaviour (correlation ≈ 0, and thus $H \approx \frac{1}{2}$) holds for periods for six months to one year. The behaviour became increasingly anti-persistent for periods of over one year.

9.4.3 *Significance*

Water storage is important for both flood control and drought tolerance. The data suggest that reservoir capacity should allow at least one year of water storage, with several years better. Naively, water storage of less than five months is likely to be inadequate because shorter dry periods are likely to be followed by similar dry periods, and similarly for wet periods. On the other hand, multiyear dry periods are likely to be followed by similarly long wet periods, and vice versa.

To make this more precise, recall the Mandelbrot (1977, 1982) formula of Section 6.2.1 relating the Hurst exponent H to the Korcak exponent B for the duration of excursions of fractal processes (on the line) above or below a given level: the probability of an excursion (flood or drought) of duration at least T scales as T^{-B}, where $H = 1 - B$. The cumulative size of the excursion scales as

$$T^{H+1}, \tag{9.4}$$

the product of its range (of the order of T^H) and duration T. In addition, larger values of B, associated with smaller values of H, correspond to relatively smaller numbers of long-lasting floods and droughts. Thus, increasing the size of a dam and thus its water storage capability is most beneficial when H is small, for example $H < \frac{1}{2}$. In the case of New Jersey data, increasing the water storage capacity of a dam from one to two years (where $H < \frac{1}{2}$) is *relatively* easier than increasing the water storage capacity of a dam from two to four months (where $H > \frac{1}{2}$).

9.5 Storage and residence times

In order to compare fractal with compartmental residence time models, consider a compartment with one input and one output. Let $f(t)$ denote the time series of input values and $g(t)$ denote the time series of output values. The compartment is said to have residence time T if $g(t)$ depends largely upon the values of f from times $t - T$ to t, and relatively little upon the values of f at times preceding time $t - T$.

In a discrete model this behaviour may be represented by the formula

$$g(t) = \sum_{s=0}^{T} a(s)f(t - s),$$
(9.5)

or more smoothly by the formula

$$g(t) = \sum_{s=0}^{\infty} \exp(-s/T)\, f(t - s)$$
(9.6)

In continuous time, one has

$$g(t) = \int_{0}^{\infty} \exp(-s/T)\, f(t - s)\, \mathrm{d}s$$
(9.7)

If the input function $f(t)$ corresponds to the increments in a Brownian function, then it is easy to see that the cumulative output

$$G(t) = \int_{0}^{T} g(s)\, \mathrm{d}s$$
(9.8)

is approximately Brownian over time scales long compared with the residence time T, but is persistent (successive increments are positively correlated, or equivalently $H > \frac{1}{2}$) over much shorter time scales. More generally, if $f(t)$ represents the increments in a fractal process $F(t)$, then $G(t)$ behaves like $F(t)$ over time scales long compared with the residence time T, but is more persistent (larger H) over much shorter time scales.

Caution. The proper formalism requires replacing $f(t - s)\, \mathrm{d}s$ by the distributional derivative of Brownian motion $\mathrm{d}B(t - s)$. The mathematics is somewhat forbidding (Mandelbrot 1965; Mandelbrot and Van Ness 1968; Mandelbrot and Wallis 1969), but it is adequate to think of the integrals as sums of many small steps.

Thus finite residence time models display several characteristic time scales, and might generate the 'multiscaling' behaviour of the time series of Sections 9.3 and 9.4, a behaviour characterized by short-term persistence, middle-term Brownian behaviour, and longer-term anti-persistence. It would be interesting to reexamine Hurst's (1951, 1956) data in this light. See also Chapter 11 below.

In contrast, Mandelbrot and co-workers showed that a compartment with an appropriate 'unbounded' residence time could transform such an input $f(t)$ into the increments of a scale-invariant process: the integral

$$g(t) = \int_0^\infty s^{H-1.5} \, \mathrm{d}B(t - s) \tag{9.9}$$

yields the increments in a fractal process with Hurst exponent H.

Part IV

Case studies

Two case studies, the first semiclassical and the second new, illustrate the science of fractal modelling. Chapter 10 unifies and reviews work of many authors in the last dozen years on pattern and process in vegetative ecosystems and concludes with a list of promising future projects ranging from studying mechanisms for succession to counting the number of species on the earth. In particular, this chapter demonstrates how fractals capture the hierarchical structure of ecosystems, in which, as first shown in the Stommel diagram, dynamics occurs on many spatial and temporal scales. Chapter 11 considers the problem of forecasting the extinction of small populations from short data sets: a crucial test for population models and a central problem in conservation biology. We use the opportunity presented by the 'case study' format to include background, supporting, and teaching material all too frequently omitted from most journal articles.

10

Case studies: pattern and process in vegetative ecosystems

10.1 Introduction

The purpose of this chapter is to show how fractal geometry might be used to measure important aspects of complex vegetation patterns, to identify scales and scaling behaviour, and to describe the underlying dynamics which gave rise to these patterns.

As background, recall Korcak's (1938) empirical search for structure in the complex distribution of areas of islands in the Aegean sea (discussed in detail in Section 3.6). Upon finding a wide range of areas, many small islands and a few large islands, Korcak applied three natural techniques to study their statistics. Areas were replaced by their logarithms in order to reduce the range of the data. The data was smoothed replacing the histogram of logarithms of areas by the corresponding cumulative frequency distribution. These steps yielded an apparently linear relationship for the number of islands of area greater than a given area a as a function of the logarithm of a. Finally, Korcak applied linear regression to find the slope of this linear relationship, yielding the power law

$$\text{number (area} > a) = \text{const} \times a^{-B}. \tag{10.1}$$

In effect, Korcak had found a fractal relationship in nature, which required Mandelbrot's discovery of fractal geometry for a full interpretation. About 20 years later, Richardson (1961) found another fractal relationship in nature: a power law for the apparent length of the coastline of England as a function of the unit 'step size' used in measurement, given by

$$\text{number of steps} = \text{const} \times (\text{step size})^{-D} \tag{10.2}$$

for a similarly fitted exponent (fractal dimension) D.

The work of Korcak, Richardson, and Mandelbrot, and apparent similarities between 'islands' of various types in vegetation maps and islands in the ocean, stimulated the first-named author (Hastings *et al.* 1982, see Section 10.2) to attempt to quantify patterns in Okefenokee Swamp vegetation with Korcak's techniques. As discussed below, many authors (cf. Bradbury *et al.* 1984; Krummel *et al.* 1987; Meltzer 1990; Sugihara and May 1990*a*) subsequently developed additional techniques and applications to ecology, using fractals to study boundaries of patches as well as their areas, and also to objectively identify natural size scales in vegetation patterns.

Fractal geometry may also contribute to the identification of ecosystem processes. Ecosystems are described as hierarchical systems, with relevant dynamics occurring on a variety of spatial and temporal scales (Allen and Starr 1982; Sugihara 1984; O'Neill *et al.* 1986; Milne 1988; May 1988). Moreover, ecosystems are open systems, with energy flowing both in and out, and typically have a large number of metastable states. One would expect in many cases that changes in the dynamics would be reflected in corresponding changes in spatial patterns and thus in the fractal exponents quantifying those patterns. In such cases sharp changes in fractal exponents computed over a limited range of scales would signal scale-dependent 'changes in the generating process with scale, and define a boundary across which one may no longer make extrapolations. In this way, fractals may provide a methodology for obtaining objective answers to such difficult problems in hierarchy theory as how to determine boundaries between hierarchical levels and how to determine the scaling rules for extrapolating within each level' (Sugihara and May 1990a).

The present chapter will address potential applications of fractals in describing scaling behaviour in vegetation patterns, and identifying certain aspects of ecosystem dynamics. In particular, we shall attempt to relate patterns in the distribution of a species or closely related group of species to their successional stage and tendency to persist. We shall also attempt to show how fractal methods can objectively determine the presence of multiple scaling regions, and thus determine natural scales in a vegetative system. Fractal geometry can thus provide objective, quantitative measure of ecosystem patterns and processes which can, at the very least, complement more complex surveys and dynamical models. Finally, we shall attempt to demonstrate the vast potential of fractals in applied ecology, in areas ranging from identifying the functional role of disturbances (shall we extinguish forest fires in a given region?) to ongoing, automated determination of the stress of grazing on heavily used grasslands.

This chapter is organized as follows. The first two sections develop the fractal geometry of vegetation patterns. The next three sections apply this geometry to models for underlying processes in vegetative ecosystems, beginning with a postulated relationship between patchiness and persistence. The last section contains a list of questions and hypotheses for further study, covering a wide variety of topics ranging from the identification of spatio-temporal scales to possible extensions to the species–area relationship.

REMARKS 10.1 Bak, Tang and Weisenfeld (1987) proposed a set of cellular automaton models for complex physical systems which give rise to fractal patterns and dynamics. These models are discussed extensively in Sections 7.2 and 8.2 above. It is interesting to speculate whether similar cellular automaton models might explain fractal patterns and dynamics in ecology. Section 10.6 develops some consequences of such a program.

10.2 Scaling behaviour

We shall begin by reviewing an early study of fractal patterns in the distribution of vegetation patches and the shapes of their boundaries (Hastings *et al.* 1982). In this work, apparent similarities between vegetation patches ('islands') in the Okefenokee Swamp and islands in the Aegean studied by Korcak suggested that Korcak's techniques be applied to search for fractal patterns in the distribution of areas of vegetation patches. Hastings *et al.* therefore attempted to fit the number of patches of area at least *a* to the hyperbolic distribution

$$\text{number (area} > a) = \text{const} \times a^{-B},$$

using a set of values of *a* spaced by semi-octaves or multiples of '*f*-stops':

$$1, \ 1.4, \ 2, \ 2.8, \ 4, \ 5.6, \ 8, \ 11.3, \ 16, \ \dots \ .$$

The Korcak patchiness exponent was determined by applying linear regression to log-transformed raw data (see Chapter 6). Results are summarized in Table 10.1. Table 10.2 presents typical raw data.

The Korcak exponent *B* measures the number of small patches relative to the number of larger patches with smaller values of *B* corresponding to fewer small patches. Classical fractal models restrict *B* to the range $\frac{1}{2} \leqslant B \leqslant 1$. Cypress distributions appeared *patchier* (more small patches) than broadleaf distributions, and cypress itself appeared patchier in the northeast than in the southeast. Informally this suggests that cypress is less persistent in the northeast, according to a postulated inverse relationship between patchi-

Table 10.1 The exponent *B* and coefficient of correlation ρ in linear regression for patches of cypress (*Taxodium ascendeus*) and broadleaf evergreen (*Magnolia virginia, Persa palustus*, and *Gordonia lasiantus*) in the Okefenokee Swamp, following Hastings *et al.* (1982)

Species, region of swamp	*B*	ρ
Cypress, northeast	0.624	0.987
Cypress, southeast	0.799	0.975
Broadleaf, northeast	0.453	0.975
Broadleaf, southeast	0.491	0.952

Table 10.2 Distribution of cypress (including mixed cypress) patches in the Okefenokee Swamp in the northeast corner (north of 34_{10} and east of 3_{70}) of the vegetation map of McCaffrey and Hamilton (1978), following Hastings *et al.* (1982). The last column is not included in Hastings *et al.* (1982), but is included here for a later critique. Eight patches smaller than 40 acres are not included due to measurement difficulties; a 40 acre patch is about $\frac{1}{4}$ inch (0.6 cm) in diameter on the 1 inch = 1 mile (1:63 360) map

Area a	Number of patches of area at least a	Number of patches of area between a and $a\sqrt{2}$
2560	3	
1810	5	2
1280	6	1
905	7	1
640	10	3
452	13	3
320	17	4
226	22	5
160	24	2
113	28	4
80	32	4
57	39	7
40	43	4

ness and persistence to be discussed in Section 10.4. Moreover, an artificial sill approximately 10 cm high built in a failed attempt to drain the swamp separates its northeast and southeast portions (Patten, personal communication). This sill partially interrupts the surface water flow, which runs predominantly from northeast to southwest. Thus, the northeast is wetter than the southeast, and thus likely a better habitat for cypress.

Broadleaf species appeared still more persistent than cypress. This is consistent with observed succession from cypress to broadleaf in the Okefenokee (Schlesinger 1978).

Conjectures. These results suggest two intriguing conjectures.

1. Late successional patterns are less patchy than early successional patterns; more precisely, the Korcak exponent *B* associated with a given type of vegetation decreases with successional stage.

2. More persistent species are less patchy than less persistent species; more precisely, the Korcak exponent *B* associated with a given type of vegetation decreases with increasing persistence.

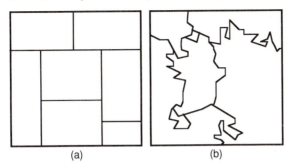

(a)   (b)

Fig. 10.1 Boundaries of artificial (a) and natural (b) patches. A sketch from Sugihara (1982). The differences can be quantified with fractal exponents (see Section 10.3, and especially Fig. 10.2).

Other exponents. As discussed in Chapter 3, many related fractal exponents might be used in attempts to find structure in complex natural forms. Krummel *et al.* (1987) used the area–perimeter exponent (equal to $D/2$) of Section 3.5 to find the fractal dimension D of irregular patch boundaries. This may be especially appropriate in cases where the boundaries arise from intense competition. Bradbury *et al.* (1984) used the dividers method of Section 3.4 to find the fractal dimension D of an Australian coral reef. As shown in Chapter 3 and Section 6.2, in many fractal models the exponents B and D are closely related by the Mandelbrot formula $D = 2B$. In particular, the fractal dimension of a patch boundary increases with increasing complexity just as the Korcak exponent increases with increasing patchiness in the distribution of areas. What does the Mandelbrot formula say about the fractal geometry of vegetation patterns? This question will be discussed in Section 10.7.1.

Sugihara (unpublished, 1982) represented this type of complexity in Fig. 10.1.

Critique. Referring to Table 10.1, note that distribution of patches by bins appears more irregular than the cumulative frequency. It is not hard to see that the expected distribution by bins should have the same scaling as the cumulative frequency:

$$\text{number}(a \leqslant \text{area} < a\sqrt{2}) = \text{const} \times a^{-B}, \qquad (10.3)$$

or, more generally, for any constant $c > 1$,

$$\text{number}(a \leqslant \text{area} < ca) = \text{const} \times a^{-B}. \qquad (10.4)$$

The use of formula 10.4 together with suitable windowing techniques should result in a better sense of fractal (or more complicated multiscaling) behaviour in the data.

Proof of formula (10.4). Fix $c > 1$. By scale-invariance,

$$\text{number}(a \leqslant \text{area} < ca) = \text{const} \times a^{-b} \qquad (10.5)$$

for some exponent b. Then

$$\text{number}(\text{area} \geqslant 1) = \text{const} \times (1 + c^{-b} + c^{-2b} + \cdots)$$

$$= \text{const}/(1 - c^{-b}), \qquad (10.6)$$

$$\text{number}(\text{area} \geqslant c^k) = \text{const}(c^{-kb} + c^{-(k+1)b} + c^{-(k+2)b} + \cdots)$$

$$= \text{const} \times c^{-kb}(1 + c^{-b} + c^{-2b} + \cdots)$$

$$= \text{const} \times c^{-kb}/(1 - c^{-b})$$

$$= c^{-kb} \times \text{number}(\text{area} \geqslant 1). \qquad (10.7)$$

Thus,

$$\text{number}(\text{area} \geqslant a) = \text{number}(\text{area} \geqslant 1) \times a^{-b} \quad (a = 1, c, c^2, ...). \quad (10.8)$$

But the power law (10.8) is characterized by the Korcak exponent $-B$. Therefore $b = B$, and formula (10.4) holds as required. $\square$

The paper by Hastings *et al.* (1982) did not consider or attempt to measure scaling regions and breakpoints separating different scaling regions. This theme was, however, soon taken up by many authors, including Bradbury *et al.* (1984), Krummel *et al.* (1987), Meltzer (1990), and Meltzer and Hastings (1992). We shall discuss their work in the next section.

10.3 Scaling regions

Fractal exponents are determined empirically as slopes of linear fits to log-transformed data. However, many authors (Bradbury *et al.* 1984; Krummel *et al.* 1987; and, more recently, Meltzer 1990 and Meltzer and Hastings 1992) found that log-transformed ecological data are best fitted using piecewise linear curves, in which the slope of each line segment is the fractal dimension over the corresponding scaling region (see Fig. 10.2).

Fractal methods thus have the potential to objectively identify scaling regions in the hierarchical construction of ecological patterns. Sugihara and May (1990a) cite the following examples.

Bradbury, Reichlet and Green (1984) investigated the fractal dimension D of boundaries of features in an Australian coral reef. They used the dividers method and rolling regressions to study whether D depends on the range of length scales. They found that D declines from approximately 1.1 at the finest scale (of the order of 10 cm) to approximately 1.05 for intermediate scales (from 20 cm to 200 cm) and rises sharply to approximately 1.15 at the largest scales (from 5 m to 10 m). [However, within each size range, D was close to constant, suggesting that scale invariant

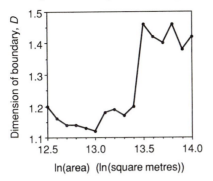

Fig. 10.2   The area–perimeter exponent over various scaling ranges, computed with a moving window technique (after Krummel *et al.* 1987). Note the sharp transition at scales of 60 to 70 ha (600 000 to 700 000 m², or 13.3 to 13.5 on the horizontal axis).

dynamics (within that size class) generated the relevant features.] These three ranges of scale correspond nicely with the scales of three major reef structures: 10 cm corresponds to the size of anatomical features within individual coral colonies (branches and convolutions); 20–200 cm corresponds to the size range of whole adult living colonies; and 5–10 m is the size range of major geomorphological structures such as groves and buttresses. That is to say, the shifts in fractal exponent at different scales appear to signal where the breakpoints occur in the hierarchical organization of reefs.

   In similar vein, Krummel *et al.* (1987) evaluated the fractal dimension of boundaries in deciduous forests in Mississippi using the area–perimeter exponent. Krummel *et al.* began with aerial photographs of the US Geological Survey (1973) Natchez Quadrangle. This region has experienced relatively recent conversion of native forests into agricultural use. The use of rolling regressions using a window of 60 points (Fig. 10.2) revealed a marked change in slope ($p < 0.001$) in the graph of log transformed perimeter and area at areas around 60–70 hectares [1 hectare = 10 000 square metres = approximatey 2.5 acres]. Small areas of forest tend to be smoother with $D = 1.20 \pm 0.02$, while larger areas, greater than 70 hectares, have more complex boundaries, $D = 1.52 \pm 0.02$. This result is interpreted to indicate that human disturbances predominate at small scales making for smoother geometry and lower D, while natural processes (e.g. geology, distributions of soil types, etc.) continue to predominate at larger scales.

Note that the dimension of 1.52 is close to that of islands $\{(x, y) : z(x, y) \geqslant 0\}$ formed by a Brownian function $z = f(x, y)$.

   Fractal analysis has also been used to measure changes in grasslands in Zimbabwe due to increasing human and cattle populations (Meltzer 1990, cf. Meltzer and Hastings 1992). Meltzer found clear evidence of two scaling regions for the cumulative frequency distribution of areas, with a single breakpoint between them. The smaller patches had a less patchy distribution than the larger. Moreover, grassland patches of size smaller than the breakpoint were less patchy (had a smaller exponent B) than patches of size larger than the breakpoint. As in Section 10.2 one may conjecture that the smaller patches are more persistent. It appeared from further analysis that large grassland patches undergo patch extinction from invasion by trees and shrubs and rapid succession. Small patches are less likely to undergo such invasion, and thus extinction of small patches follows another route.

Critique. The cumulative frequency distribution counts all patches in determining the Korcak exponent B from the number of patches of size greater than a. The process of forming the cumulative frequency distribution may thus cross scale boundaries in systems with several scaling regions. In such systems the histogram of frequencies of areas (see Table 10.2) should be used instead of the cumulative frequency distribution. Formula 10.5 above justifies this approach.

REMARKS 10.2 There are several ways to ascertain the existence of multiple scaling ranges separated by breakpoints. First, one can simply plot log-transformed data points, and decide whether they appear to lie on a straight line. This simple visual test can be supplemented by appropriate statistical tests such as the Durbin–Watson test (Section 6.6; cf. Draper and Smith 1981). These tests measure the distribution of residuals, that is, displacements of data points above and below the regression line. A linear model is appropriate if the residuals are random and independent.

Finally, regression can be restricted to subsets of the data in order to determine local slopes over these subsets. One approach is to perform regression over a window containing a fixed number of data points, and let the window range over the entire set (see Krummel *et al.* 1987). This method is called rolling regression. Alternatively, one can postulate the existence of one or more breakpoints, and perform regression separately in each scaling range (see Meltzer 1990; Meltzer and Hastings 1992). If the computations over two adjacent windows (overlapping only at their common endpoint) give apparently distinct regression slopes, and thus exponents, one can then determine whether the slopes are significantly different. It is also easy to vary the breakpoint between adjacent windows and thus minimize the overall mean square error.

10.4 Patchiness and persistence

We have seen that fractal geometry can be used to objectively identify scaling regions in vegetative ecosystems and scaling behaviour within each scaling region. We have also postulated an inverse relationship between patchiness and persistence in vegetative ecosystems, based on field data from the Okefenokee Swamp (Section 10.2). Sugihara and May (1990a) suggested the possibility of a more general inverse relationship between patchiness and persistence, based upon observations on a variety of different systems: satellite ocean colour data and patch dynamics of bryozoan and coral colonies (Jackson and Hughes 1985).

Thus fractal geometry appears to offer the promise of forecasting persistence in vegetative ecosystems from a single set of measurements of spatial patterns. We shall explore this question theoretically with a variety of mathematical models in the next two sections.

10.5 Model building

Fractals can be used to compare patterns generated by postulated dynamical models with patterns in nature. In this section we shall review some of the history of ecosystem models in space and time, and show how fractals can be used to assess such models, following Hastings *et al.* (1982).

Levin and Paine (1974) developed patch dynamics in order to model immigration and extinction in space and time. They represented space as a two-dimensional grid of cells (we use the term 'cell' rather than their term 'patch' in order to reserve the term 'patch' for a connected region of cells of one type). The cells are assumed to be so small that each cell may be assigned a unique state from a finite set of states. The states of all cells are updated by deterministic or probabilistic rules at discrete time intervals (for a chosen small time step Δt, *assumed short compared with all natural time scales*, so that the dynamics is essentially continuous). Usually, and unless otherwise stated, the state of the system at time $t + \Delta t$ depends only on its state at time t, that is, the system is *Markovian*.

The dynamics of immigration and extinction are captured in a toy system with just two competing species, A and B, and thus three states: those occupied by species A (denoted simply A), those occupied by species B (denoted simply B), and those that are vacant. Extinction is modelled by a Markovian process: cells of type A have a characteristic mean lifetime L_A, and cells of type B have a characteristic mean lifetime L_B. Let $r_A = 1/L_A$ be the cell extinction rate. Then, in time Δt, a fraction $r_A \Delta t$ of the cells occupied by species A will become vacant. Extinction of cells of type B is modelled similarly. Suppose also that vacant cells are immediately recolonized.

One simple mean field model (Levin and Paine 1974; see Hastings *et al.* 1982) assumes that the probability that a cell is recolonized by species A is proportional to the product of number of cells of type A and a characteristic *diffusion rate* D_A, and similarly for species B. We call this model a mean field model since propagules can come from throughout the system. If a fraction x of all the cells is currently occupied by species A, then the fraction

$$xD_A/[xD_A + (1 - x)D_B] \qquad (10.9)$$

of vacant cells is colonized by species A. The rest of the vacant patches are

Table 10.3   State transitions in the toy mean field model

Transition	Fraction of all cells undergoing transition within time Δt
A → vacant	$r_A x \, \Delta t$
B → vacant	$r_B(1-x) \, \Delta t$
A or B → vacant	$[r_A x + r_B(1-x) \, \Delta t$
A or B → vacant → A	$[r_A x + r_B(1-x)][x D_A/(x D_A + (1-x) D_B)] \, \Delta t$
A or B → vacant → B	$[r_A x + r_B(1-x)][(1-x) D_B/(x D_A + (1-x) D_B)] \, \Delta t$

colonized by species B. Table 10.3, taken from Hastings *et al.* (1982) summarizes the state transitions.

The dynamics of this model are easily derived, again following Hastings *et al.* (1982). Subtracting the cell extinction rate from the colonization rate yields

$$\Delta x/\Delta t = -r_A x + [r_A x + r_B(1-x)]\{x D_A/[x D_A + (1-x)D_B]\}. \quad (10.10)$$

Formula (10.10) can be simplified by dividing by x, obtaining the logarithmic rate of change

$$(1/x)\,\Delta x/\Delta t = -r_A + [r_A x + r_B(1-x)]\{D_A/[x D_A + (1-x)D_B]\}. \quad (10.11)$$

Additional calculations yield

$$(1/x)\,\Delta x/\Delta t = \{-r_A[x D_A + (1-x)D_B] + [r_A x + r_B(1-x)]D_A\}$$
$$\times [x D_A + (1-x)D_B]^{-1}$$
$$= (-r_A D_A x - r_A D_B + r_A D_B x + r_A D_A x + r_B D_A - r_B D_A x)$$
$$\times [x D_A + (1-x)D_B]^{-1}$$

(after writing the numerator as a sum of terms)

$$= (-r_A D_B + r_B D_A)/[x D_A + (1-x)D_B] \qquad \text{(after cancellation)}.$$
$$(10.12)$$

If the ratios of diffusion to extinction rates satisfy the condition

$$D_A/r_A > D_B/r_B, \qquad (10.13)$$

then $(1/x)\,\Delta x/\Delta t > 0$, and the fraction of patches x occupied by species A increases to 1, that is, species A displaces species B.

Even this toy model has interesting behaviour. First, the long-term dynamics of any single species depends only upon the diffusion–extinction ratio D/r. However, if species B has a faster diffusion rate, a vacant region adjacent to a region containing both species would initially be colonized by species B, and later go over to species A. There are thus two distinct time regimes (Levin and Paine 1974).

It is easy to construct similar but more realistic models. For example, the diffusion rates may be interpreted as the effective areas over which propagules spread from individual cells. That is, suppose that propagules of species A travel a characteristic distance s_A and propagules of species B travel a characteristic distance s_B, each in a unit time step. In a mathematical sense, the diffusion rate is the ratio $\Delta s^2/\Delta t$, and thus

$$D_A = s_A^2, \qquad D_B = s_B^2, \qquad (10.14)$$

There is also a simple geometric interpretation. Propagules from a cell occupied by species A typically travel over a circle of radius s_A centred on that cell, and thus cover an area πs_A^2, and similarly for species B. As in the mean field model, suppose that vacant cells are colonized by each species with probability proportional to the number of propagules of that species arriving there. Then the probability that a given cell is colonized by species A is given by the formula

$$\pi s_A^2 x/(\pi s_A^2 x + \pi s_B^2 y) = s_A^2 x/(s_A^2 x + s_B^2 y), \qquad (10.15)$$

where x is the fraction of cells within a circle of radius s_A occupied by species A, and y is the fraction of cells within a circle of radius s_B occupied by species B. Thus, by analogy with formula (10.9), the effective diffusion rates are $D_A = s_A^2$ and $D_B = s_B^2$, respectively. Note that y is not necessarily equal to $1 - x$ unless $s_A = s_B$. In that case, species A drives out species B under the same conditions as in the mean field model.

However, the analysis is already more complex. In general, the centres of large patches of one type are not invaded, but for a patch of type A, a *boundary layer* (neighbourhood of the boundary) of characteristic width s_B can be invaded.

One can also consider history-dependent (non-Markovian) models, in which the fate of each cell depends not only upon its current state but upon its history of occupancy. The use of non-Markovian models is likely to be more appropriate since they include the effect of the history of cell occupancy upon the probability of cell extinction. However, the analysis becomes significantly more complex and one must turn to simulation methods, as in Section 7.2.

We conclude by considering a possible role of fires in regulating the classical succession of pine or other conifers to broadleaf forests. As hypothesized, the rapid spread of pines corresponds to a faster effective diffusion rate for pines. Because shade is more beneficial to broadleaf germination, and broadleaf trees overtop pine trees, broadleaf trees should exhibit reduced extinction and therefore a greater diffusion–extinction ratio. In the absence of fires, succession should continue from pine to broadleaf as the diffusion–extinction ratio drives the dynamics. In the presence of frequent fires, succession is interrupted as the fires periodically 'reset the clock to zero', and keep the system in its original diffusion-oriented regime, and pines with faster effective diffusion should predominate.

Project. Study these models with cellular automata simulations as discussed in Section 7.2. How do the fractal exponents depend upon the choice of model or simplifying assumptions used? Does each model have a distinct set of exponents?

REMARKS 10.3 This model predicts that the most persistent species (as measured by patchiness) is also the most common at a successional climax. It would be interesting to test this hypothesis; see Section 10.7.2.

10.6 A possible formalism for patchiness and persistence: evolution of CA models in time and space

As a thought-experiment, consider time lapse photography of a forest. This yields a sequence of vegetation maps at regular time intervals. Consider these maps as slices in a map showing the evolution of vegetation in time and space, analogous to the representation of the evolution of cellular automata in time and space (see Section 7.2).

Under an *assumed* self-similar fractal behaviour in space and time together, there is a simple relationship between the dimensions D_{time} of the state transitions of a cell in time (see Application on p. 54; Mandelbrot 1977, 1982; Vicsek 1989; Erzan and Sinha 1991) and the dimension D_{space} of cell boundaries in space (the dimension D computed above, the subscript 'space' is used for emphasis):

$$D_{time} = D_{space} - 1. \tag{10.16}$$

Since both fractal exponents D_{time} and D_{space} increase with increasing irregularity, formula (10.16) associates irregular temporal distributions (lack of persistence) with irregular spatial distributions. Moreover, the time dimension of transitions of a fractal (generalized Brownian) process is related to the scaling exponent H for that process by the Mandelbrot (1977, 1982) formula (see Sections 3.6 and 6.2):

$$H - 1 - D_{time} \tag{10.17}$$

Combining formulae (10.16) and (10.17) yields the postulated inverse relationship between persistence H and patchiness D ($=D_{space}$):

$$H = 2 - D. \tag{10.18}$$

The simple Markovian model has $H = 0.5$, and thus $D = 1.5$, close to the value of 1.52 in Krummel *et al.*'s (1987) study of deciduous forests.

Critique. Observed *B* values outside the range $0 < B < 1$ implied by formula (10.18) and the relationship $D = 2B$, suggesting that a more complex multi-species analysis may be needed for real systems (Meltzer 1990; cf. Meltzer and Hastings 1992). Nonetheless, ecological data, thought-experiments, and simulation analysis suggest the possible ubiquity of a qualitative inverse relationship between the patchiness and persistence.

10.7 Future projects

This chapter sketches several speculative applications of fractals to a variety of questions in ecology, ranging from identification of dynamical scales in space and time, motivated by the Stommel diagram (Stommel 1963, 1965; Haury *et al.* 1978) to the problem of counting the number of species on the earth. We begin with several questions related to succession.

10.7.1 *Mechanisms for succession*

At least in principle, fractals can be used to study whether succession is determined by invasion or senescence, and to study the role of boundaries in competition. Many processes have characteristic or universal fractal exponents, which can be compared with field data in order to test their applicability.

For example, our Markovian model predicts that boundaries have fractal dimension $D = 1.5$. The physics of fluids in porous media may provide another model for competition along boundaries in which one species actively displaces another. In this model, viscous fingering (cf. LeNormand 1989; Mandelbrot 1982), and, in a related model, diffusion-limited aggregation (cf. Meakin and Tolman 1989; Mandelbrot 1982; see our Sections 8.3 and 8.4 above), boundaries have fractal dimension $D = 1.7$.

The existence of boundaries which are too regular to arise from Markovian extinction and random short-range immigration might be explained in many ways, instead of *or* in addition to non-Markovian extinction rules. However, non-Markovian extinction rules remain an attractive possibility in terms of known species–environment interactions.

For example, oak leaf litter is a natural fire retardant, and thus reduces the probability that a cell occupied by oaks becomes vacant from fire. Thus, regions dominated by oak trees might have relatively smooth boundaries, with fractal dimension close to 1.

Cypress bark contains tannin, making water in cypress swamps acid, and inhibiting invasion by competitors. This might again leave few holes and relatively smooth boundaries. What is actually observed? In fact (see Section 10.2) cypress patches in the wetter northeast part of the Okefenokee Swamp had Korcak exponent $B = 0.624$. This corresponds to D approximately 1.2 by the Mandelbrot formula $D = 2B$ (see Sections 3.6 and 6.2). Mixed broadleaf (*not* oak) patches had $B < 0.5$, corresponding to D approximately 1.

In contrast, Krummel *et al.* (1987) found D approximately 1.5 from measurements of the boundaries of large patches in a disturbed forest environment, as predicted by the Markovian model.

10.7.2 *Testable hypotheses*

Here is a short list of hypotheses which should be tested in further work on fractal models.

1. Patchiness (as measured by the exponent B) decreases with successional stage. If true, this would provide an objective test for successional stage, and evidence as to whether a system has reached a climax state—in a climax state the species with the least patchy distribution should be the most abundant. This hypothesis is suggested by both the Okefenokee data (Hastings *et al.* 1982) discussed above and the intuitive idea of succession moving toward a stable climax state in the absence of disturbances. Moreover, if this hypothesis is true, then fractal methods might be used to help select indicator species, which would be most susceptible to ecosystem changes.

2. Regions where disturbance plays a major role in patterns can be identified by comparing the area coverage with the patchiness. If the dominant species, as measured by area coverage, have more patchy distributions than other species, then disturbances play a role in vegetation patterns, and conversely. These ideas might be used in forest and crop management. In particular, consider an ecosystem which is subject to relatively regular fires, such as Yellowstone National Park. There has been intense debate about whether to extinguish all fires, particularly destructive fires, those fires which affect nearby communities, or no fires. It is clearly important to obtain objective measures of the role of fires in preserving the ecosystem diversity, for example by holding back succession. See also Section 10.7.1 above.

3. Early species in *senescence-mediated succession* have less patchy distributions than comparable species in *overcropping mediated succession*. See also Section 10.7.1 above.

10.7.3 *Discussion*

It is easy to find dynamics which explains the existence of the two scaling regions, as in Section 10.3. One need merely study grassland-shrub-try systems and postulate relatively slow long-range immigration of shrubs and trees, together with rapid succession of grassland to shrubs and trees. Suppose further that the probability that a given patch is invaded is proportional to its area. Then the succession process causes 'cell extinction' in larger patches. The small patches are not similarly subject to succession, and their cell extinction process is determined by nearest-neighbour interactions with adjacent vacant land and other types.

The study of parasites and plant diseases might provide another example. Consider a hypothetical parasite or disease which is maintained at low endemic levels, spreads randomly, and rapidly destroys patches of a particular species; examples include Dutch elm disease. Such a parasite or disease would destroy large patches much more often than small isolated stands.

It would be interesting and important to study the role of diversity in regulating agricultural pests.

10.7.4 *Measurement of dynamical scales*

The Stommel diagram (Stommel 1963, 1965; Haury *et al.* 1978) is a fascinating representation of the many spatial and temporal scales of variability in the dynamics of ocean plankton. We list here the predominant scales associated with various levels in the Stommel diagram and derive implicit spatio-temporal exponents H, which have implications for the consequent underlying dynamics. The exponent H is given by the renormalization relationship

$$\Delta s^2 = \Delta t^{2H} \tag{10.19}$$

of fractal processes (see Section 2.4). Consider a process on temporal scales $T_1 < t < T_2$ and spatial scales $S_1 < s < S_2$. Formula (10.19) implies that

$$(S_2/S_1)^2 = (T_2/T_1)^{2H}. \tag{10.20}$$

We convert formula (10.20) to a formula for 'orders of magnitude' by applying base-10 logarithms:

$$\log_{10}(S_2/S_1)^2 = \log_{10}(T_2/T_1)^{2H},$$

and thus

$$\log_{10} S_2 - \log_{10} S_1 = H(\log_{10} T_2 - \log_{10} T_1).$$

Solving for H yields

$$H = (\log_{10} T_2 - \log_{10} T_1)/(\log_{10} S_2 - \log_{10} S_1), \tag{10.21}$$

the quotient of the number of orders of magnitude in the temporal scales divided by the number of orders of magnitude in the spatial scales.

Most of the processes in Table 10.4 display scaling exponents of approximately 0.5, the value in diffusion processes, or 1.0, the value in deterministic transport processes. There are two exceptions. Swarms may display large-scale spatial coherence, and the apparent exponent $H = 3$ appears to reflect this coherence. The value $H = 0.3$ for small ocean basins probably reflects physical boundaries and their role in making spatial variability grow

Table 10.4 Spatial and temporal scales and the number of orders of magnitude represented in each scale, derived from the Stommel diagram. The associated fractal exponent H is obtained by dividing the number of spatial orders of magnitude by the number of temporal orders of magnitude. The apparent value of $H = 3$ for swarms is not meaningful, and is explained below

Hierarchical level	Temporal scale		Spatial scale		H
	Range	Orders of magnitude	Range	Orders of magnitude	
A. Micro patches	4 to 330 min.	1.8	0.3 to 3 m	1.0	0.6
B. Swarms	1.5 to 6 days	0.5	60 to 1600 m	1.4	3.0
C. Upwellings	30 to 300 days	2.0	5 to 1600 km	2.5	1.2
D. Eddies and rings	1 to 10 yrs.	1.0	60 to 600 km	1.0	1.0
E. Island effects	0.03 to 3000 yrs.	5.0	0.16 to 100 km	2.8	0.6
F. El Nino type events	10 to 100 yrs.	1.0	160 to 1600 km	1.0	1.0
G. Small ocean basins	50 to 3000 yrs.	1.8	100 to 300 km	0.5	0.3
H. Biogeographic prov.	300 to 10000 yrs.	1.5	600 to 6000 km	1.0	0.6

Spatial scales for reference:

K. Width of oceanic fronts — 1 km
J. Width of oceanic currents — 60 km
I. Length of oceanic currents and fronts — 600 km

more slowly than a classical Brownian process. We can only speculate about this type of analysis on the basis of the Stommel diagram, but future possibilities seem exciting. See also Frontier (1987).

10.7.5 *Automated measurement of ecosystem patterns and processes*

Recent advances in remote sensing and vegetation mapping offer the intriguing possibility of inexpensive automated surveys to detect early changes in ecosystem patterns and processes (Meltzer 1990; Meltzer and Hastings 1992).

Satellite images are typically available with 30-metre resolution, significantly better than the resolution used by the authors cited here. Tilton (1987) described a computerized contextual classifier to produce vegetation maps automatically from multispectral satellite data. The advent of more powerful parallel computers will make computerized classification easy and readily available. Computerized classification offers several advantages over hand classification—it is easier to standardize, there is no artificial smoothing of boundaries, and small patches will not be missed.

Finally, it is easier to directly analyse digital images than 'analogue' vegetation maps which involve a second translation back to digitized data in order to compute fractal exponents.

It is estimated that with these advances, one person with an 80386 SX based personal computer could analyse surveys of typical grasslands several times per year, and thus provide easily affordable backup to more expensive field ecologists. Finally computerized classification has the advantage of being objective, not subjective.

10.7.6 *Model verification with the use of several computations of exponents*

As in Section 10.2, Korcak's (1938) patchiness exponent B has been widely used to study vegetative ecosystems for several reasons. First, B directly measures patchiness as defined by a distribution of areas. Secondly, exponents involving perimeters involve accurately knowing the boundaries of patches, and accurately measuring their perimeters. Finally, there is a conjectured connection between patchiness and persistence (Sections 10.2 and 10.4–10.6).

On the other hand, it may be more appropriate to use fractals for studying boundary irregularities. For example, Bradbury *et al.* (1984) used the 'dividers' method in another setting, and Krummel *et al.* (1987) used the area–perimeter exponent (see Section 10.3). Box counting techniques can also be used to determine the fractal dimension of the patch boundaries (see Section 3.5; see also Frontier (1987)).

The computation of several fractal exponents in a single application might be a useful way to further test the applicability of fractal models. For example, in a fractal set of islands, the fractal dimension of the boundary D is twice the Korcak exponent B (see Section 3.6.1).

10.7.7 *Counting species on earth*

The species–area exponent arises in a well-known relationship (Preston 1962; MacArthur and Wilson 1967; May 1975; Sugihara 1980) between the number s of species found in a sampling region, its area A, and linear scale l:

$$s = \text{const} \times A^{0.25} = \text{const} \times l^{0.5}. \tag{10.22}$$

Consider now searching for new species in an expanding family of circles centred at a given point. Let X be the set of points where new species are found; that is, moving out from the centre into larger circles, a point is added to X each time a new species is found. By the above formula, X has cluster dimension 0.5 (see Section 3.3), and thus space appears 0.5-dimensional from the point of searching for new species. Since new species would then appear on a Cantor-like set of dimension 0.5, we need a search space of dimension more than $2 - 0.5 = 1.5$ (by the intersection formula of Section 3.7.3 the search space will meet X in a set of positive dimension, (cf. Lovejoy *et al.* 1986)).

In addition, the telescope–microscope formula of Section 3.1 relates the species–area formula to the effects of reaching down to look for smaller and smaller species. Suppose we have found s species of size at least Δl in a quadrat of linear scale l, and thus a relative scale $l/\Delta l$. By the species–area formula, we expect approximately $k^{0.5}s$ species of size at least Δl in a quadrat of linear scale kl or relative scale $kl/\Delta l$. However, searching for even smaller species, of minimum size $\Delta l/k$, on the original quadrat also increases the relative scale to $kl/\Delta l$, and yields an expected number of species $k^{0.5}s$. Fractals thus suggest, *as a testable hypothesis*, an extended species–area relationship

$$s = \text{const} \times (l/\Delta l)^{0.5}. \tag{10.23}$$

However, May's (1978, 1988, 1990) work on the number of species shows that formula (10.23) *cannot hold on all scales*. 'Very roughly, as one goes from animals whose characteristic linear dimension [Δl] is a few metres down to those of around 1 cm (a range spanning many orders-of-magnitude in body weight), there is an approximate empirical rule which says that for each tenfold reduction in length (1000-fold) reduction in body weight) there are 100 times the number of species.' (May 1990, p. 178). Thus s scales inversely as Δl^2, and not as $\Delta l^{0.5}$.

There may be several explanations. First, Hutchinson and MacArthur (1959) argued that the number of species should scale with the number of new roles (niche hypervolume) and that terrestrial organisms, which see the world as two-dimensional, should see a two-dimensional niche space. On the scale of Δl, this argues for the empirical rule that s scales inversely as Δs^2. This argument, however, suggests a species–area exponent of 0.5. May (1988) applied the hypothesis of Morse *et al.* (1985) that roughly equal amounts of energy flowed through each size category (cf. Odum 1953). Since the mass of an animal scales as the cube of its size, and its energy consumption as the 0.75 power of its mass (Peters 1983), the number of individuals should scale approximately as $\Delta l^{-2.25}$. Moreover, work of Morse *et al.* (1985) on availability of space on various types of vegetation found that leaf boundaries had dimension approximately 1.5, and thus that leaves have dimension approximately 2.5.

These observations suggest that a possible sharp breakpoint in the scaling behaviour of the number of species s as a function of the linear scale l. On scales of 10 metres and above, s scales as $l^{0.5}$, in accord with the usual species–area formula (10.22). However, on scales of 1 metre and below, s scales as l^2. In order to reconcile these observations, we suggest the following curious hypothesis: the niche space present in a habitat of spatial scale l had dimension 2 for scales l of order 10 metres and above, but has *dimension 8 on scales l* of order 1 metre and below. Thus niche volume is proportional to l^2 for scales l of order 10 metres and above and proportional to l^8 on scales of 1 metre and below.

Substituting these formulae for the niche volume into the generalized species–niche volume relationship

$$n = \text{const} \times (\text{niche volume})^{0.25}, \tag{10.24}$$

as described by Sugihara (1980) yields the predicted scaling behaviour of the number of species, both above and below the predicted breakpoint. The high dimension of niche space at small scales may reflect both habitat volume (Morse *et al.* 1985; May 1990; Scheuring 1991), and specialization and competition in several nonspatial dimensions (cf. May 1990). In particular, Scheuring argues that the habitat dimension is the fractal dimension of the vegetation (Morse *et al.* 1985), which is typically strictly between 2 and 3.

Case study: scaling behaviour of density-dependent populations under random noise

11.1 Introduction

The prediction of local species extinction is a useful goal and test for population models. Essentially by definition, the population levels of a species which persists for a long time remain within a bounded range, and thus there is an asymptotic value to the range of population fluctuations. The challenge is therefore to predict the long-term range of population values from available time series which are frequently relatively short. (The range of a population over a time interval is defined to be the difference between the maximum and minimum population values.) Sugihara and May (1990a) suggested that the range of population values might sometimes follow fractal or power law scaling: over a time interval of duration Δt, the range of population values $R(\Delta t)$ scales as

$$R(\Delta t) = \text{const} \times (\Delta t)^H. \tag{11.1}$$

In this case, the scaling exponent H measures the rate of growth of population fluctuations with the time interval Δt. Larger values of H correspond to more rapid increases of the range of fluctuations. Thus, neglecting the role of the constant in equation (11.1) and initial population size, larger values of H should correspond to increasing likelihood of extinction. A preliminary investigation (Sugihara and May, 1990a) of two bird species, the least flycatcher and the American redstart, appeared to confirm this conjecture.

The use of fractals to study local extinction is appealing; however, several important issues remain. The first problem involved the large degree of uncertainty in estimating H from the field data, particularly from time series which typically contained 10 to 15 data points. The second problem involved model verification. Is the underlying population dynamics truly fractal? Finally, can this approach be extended across a wide spectrum of bird species?

This chapter addresses these problems. Some technical problems in computing fractal exponents from the given field data are overcome— however, the uncertainty in computed exponents remains relatively large because the time series are very short. Theoretical and simulation methods confirm the existence of this uncertainty.

We also consider two alternative approaches to modelling the dynamics of the populations under study. The first approach involves the nonlinear prediction techniques of May and Sugihara (1990b). These methods appear to offer the promise of a multispecies model using only the time series of population values of a single species. The first step in applying these methods is to compute the dimension of the dynamics of the system (Grassberger and Procaccia 1983). However, it takes the order of 10^D data points to identify a D-dimensional system (Smith 1988; Ruelle 1990; Ghil *et al.* 1991), and the time series under study contain only 10–30 data points. Thus nonlinear methods are not practical in this study.

The second approach involves linear density-dependent models which make use of the hypothesis that fluctuations of natural populations depend upon both environmental noise and density-dependent control. The scaling behaviour of the density-dependent model is examined and contrasted with the scaling behaviour of the fractal model.

In particular, note that the range of population values of any species with arbitrarily long-term persistence is necessarily bounded. Thus the fluctuations cannot follow a single power law (11.1) since that would imply an unbounded range of fluctuations. Moreover, samples of a bounded population with some random fluctuations taken at sufficiently long time intervals are close to uncorrelated. Thus, over such intervals, applying the local growth of moment method for computing fractal exponents (Section 4.3) yields

$$\rho = E((x_2 - x_1)(x_1 - x_0))/E((x_1 - x_0)^2) = -E(x_1)^2/2E(x_0)^2 = -\tfrac{1}{2}, \quad (11.2)$$

and thus $H = -1$. (Long-term persistence implies that the mean population increment is 0, in which case ρ may be interpreted as the coefficient of correlation.)

Both the alternative model and field data are found to exhibit more than one scaling region, with the scaling exponent a decreasing function of the sampling interval, in contrast to the power law (11.1).

We compare these models by applying them not only to natural populations but also to simulations. Simulated data is generated using both random walk models (the simplest fractal models) and the density-dependent model developed below. The fractal exponent H associated with each data set is computed in several ways, as in Chapter 4. One would expect each of these methods to yield the same value of H for fractal processes. Other tests for fractal behaviour (Section 6.5) are also applied to the data. Unfortunately, many results are inconclusive or limited by the relative shortness of population time series. Nonetheless, computation of fractal exponents is shown to be useful in understanding the scaling behaviour of population fluctuations.

This chapter is organized as follows. Linear density-dependent models are described in Section 11.2, and their scaling behaviour is derived in Section 11.3. We next compare all three models: nonlinear prediction, fractal, and linear density dependent, beginning with general considerations in Section 11.4. Tests for comparing the other two models are developed in Section 11.5. Results occupy the next two sections, followed by a discussion in Section 11.8. We used bird population data from Holmes *et al.* (1986), and butterfly population data from Harrison *et al.* (1991) for the first 27 years of data and Ehrlich (personal communication) for the last 2 years of data; see also Ehrlich (1965), Ehrlich and Murphy (1981), Harrison *et al.* (1986), and Murphy *et al.* (1986). We invite readers to try other techniques for analysing these data. The bird data are given in the Appendix at the end of this chapter.

11.2 A linear model for density dependence

We shall begin with a linear model for density dependence with additive noise because such a model is easy to analyse mathematically as well as by simulation. We shall also see that our conclusions hold *qualitatively* for nonlinear models displaying density dependence. The linear model is obtained by first linearizing the well-studied logistic equation (in continuous time) about its stable equilibrium. The logistic equation

$$\frac{dy}{dt} = ry(1 - y/K) \tag{11.3}$$

is perhaps the simplest differential equation exhibiting density dependence, that is, a decrease in the rate of population growth at large populations. The logistic model is parametrized by an intrinsic growth rate (the growth rate for very small population levels) r, and a carrying capacity K. It is easy to see that the logistic equation has a stable equilibrium at $y = K$, and that the linearized system about this equilibrium takes the simple form

$$\frac{dx}{dt} = -rx, \tag{11.4}$$

where $x = y - K$, and $-r$ is the eigenvalue or *Lyapunov exponent*. Equation (11.4) may be considered as a prototype for density dependence in that there is a stable equilibrium population level, and the return of the population level to equilibrium is characterized by a linear differential equation. The effect of random fluctuations is easily introduced by rewriting equation (11.4) in differential form and adding a random walk term dB to yield the *linear density-dependent* model

$$dx = -rx\,dt + dB. \tag{11.5}$$

For later reference, the linear density-dependent model has a *natural time scale*

$$t_0 = 1/r; \tag{11.6}$$

a fluctuation decays to a fraction $1/e$ or about 37% of its original value after a time t_0.

This model may be used to generate time series of simulated population fluctuations whose statistics can be readily compared with those of natural populations. It is worth noting that equation (11.4) represents a linear model with additive noise despite the nonlinearity of the logistic equation. The scaling behaviour of model populations governed by equation (11.5) can be determined analytically. In contrast, determining the scaling behaviour of more complex models would appear to require a combination of qualitative results and simulation methodology. We shall see below that the scaling behaviour of model (11.5) is typical of that of more complex models, and thus that our conclusions are more robust than many conclusions drawn from linear models.

We now compare the linear density-dependent model to fractal models used by Sugihara and May (1990a). First, recall that fractal processes are stationary. It is easy to see that time series generated by the linear density-dependent model are asymptotically stationary since the effect of the initial conditions, or any subsequent fluctuations, decays as $\exp(-rt)$. Consequently, the distribution of $\{x(t)\}$ for large t can be found by integrating

$$\int_0^t \exp[-r(t-s)] \, dB(s). \tag{11.7}$$

Thus, for large t, the distribution of $\{x(t)\}$ is asymptotically normal with mean 0, just as for the fractal model.

However, in contrast to the case of fractal processes, the increments Δx depend explicitly upon the value of x. This can be used to decide whether data are better modelled by fractal or linear density-dependent models.

REMARKS 11.1 The linear density-dependent model (11.5) is a one-parameter generalization of random walks in the sense that setting the intrinsic growth rate r equal to 0 yields a random walk.

We now derive the scaling behaviour of the linear density-dependent population model with additive noise.

11.3 The main theorem

THEOREM 11.1 Consider the linear density-dependent population model with additive noise (equation (11.5) above):

$$dx = -rx \, dt + dB.$$

In the limit of large t, for any time step h, the increments $\Delta x(t) = x(t + h) - x(t)$ have expectation 0, and successive increments $\Delta x(t) = x(t + h) - x(t)$ and $\Delta x(t + h) = x(t + 2h) - x(t + h)$ have coefficient of correlation $\rho = -\frac{1}{2}[1 - \exp(-rh)]$.

Proof. We begin with some preliminary observations. Since t is large, we are in the asymptotically stationary region and need not consider initial conditions. Thus, the expectation of any increment is 0. Since the model equation (11.5) is linear, we may obtain $x(t + h)$ from $x(t)$ by first applying the noise-free equation

$$dx = -rx \, dt \qquad (11.8)$$

(which, incidentally, describes the mean process associated with equation (11.5)) and then adding the effect of noise in the time interval from t to $t + h$. We obtain

$$x(t + h) = e^{-rh}x(t) + \Delta w', \qquad (11.9)$$

$$x(t + 2h) = e^{-rh}x(t + h) + \Delta w''$$

$$= e^{-2rh}x(t) + e^{-rh}\,\Delta w' + \Delta w''. \qquad (11.10)$$

Here $\Delta w'$ represents the effect of the random walk term $dB(s)$ over the time interval $t < s < t + h$. We have

$$\Delta w' = \int_t^{t+h} \exp[-r(t + h - s)] \, dB(s), \qquad (11.11)$$

and $\Delta w''$ is given by a similar integral over the time interval $t + h < s < t + 2h$. Since the process $\{x(t)\}$ is asymptotically stationary, that is, its statistics are independent of t for large t, we may write

$$\sigma^2 = E([x(t)]^2) = E([x(t + h)]^2). \qquad (11.12)$$

Similarly, using stationarity of the random walk process:

$$\sigma^2_{\Delta w} = E((\Delta w')^2) = E((\Delta w'')^2). \qquad (11.13)$$

We now compute

$$\sigma^2 = E([x(t + h)]^2)$$

$$= e^{-2rh} E([x(t)]^2) + 2E(e^{-rh}x(t)\Delta w') + E((\Delta w')^2)$$

(by equation (11.9))

$$= e^{-2rh} E([x(t)]^2) + E((\Delta w')^2)$$

(since $x(t)$ and $\Delta w'$ are independent and $E(\Delta w') = 0$)

$$= e^{-2rh}s^2 + \sigma^2_{\Delta w}.$$

Thus, simplifying the above calculation, we have

$$\sigma^2 = e^{-2rh}\sigma^2 + \sigma_{\Delta w}^2.$$

A series of straightforward calculations now yields the variance

$$E([x(t + h) - x(t)]^2) = 2\sigma^2(1 - e^{-rh}), \qquad (11.14)$$

the same result for the variance of the next successive increment, and the covariance

$$E([x(t + h) - x(t)][x(t + 2h) - x(t + h)]) = -\sigma^2(1 - e^{-rh})^2. \quad (11.15)$$

Combining equations (11.14) and (11.15) yields the required formula for the correlation of successive increments

$$\rho = -\tfrac{1}{2}(1 - e^{-rh}). \qquad (11.16)$$

$\square$

REMARKS 11.2
1. The same result holds for the corresponding discrete-time model, obtained by discretizing with time step τ, with a stable equilibrium with eigenvalue $\exp(-r\tau)$, provided that the time step τ is sufficiently small that the discretization closely follows the differential equation. (Instabilities may result at large time steps compared with the intrinsic time scale $1/r$; see Section 7.4). The proof is similar and omitted.
2. The amplitude of $\{x(t)\}$ (for example, the square root of the variance of $\{x(t)\}$) depends linearly upon the amplitude of the noise, once the process is in the asymptotically stationary range, that is, for time t much larger than the natural time scale $t_0 = 1/r$). The amplitudes of the increments depend similarly upon the amplitude of the noise. Both statements are easy consequences of formula (11.7). However, formula (11.14) provides the most useful estimates. Since the population increments have expectation 0, their variance is the same as their second moment. The variance of a typical increment over one time step, σ_1^2, and the variance of the difference between the population and its mean value, σ^2, are related as follows:

$$\sigma_1^2 = 2\sigma^2(1 - e^{-r}). \qquad (11.17)$$

More generally, over h time steps, the relationship is

$$\sigma_h^2 = 2\sigma^2(1 - e^{-rh}). \qquad (11.18)$$

The linear model with additive noise has a *natural time scale* of $1/r$ in contrast to fractal models. This yields three broad but distinct scaling regions, depending upon whether the time step under consideration is much shorter than $1/r$, approximately equal to $1/r$, or much larger than $1/r$, each with its own approximate power law. Table 11.1 summarizes the power laws. In particular, this model predicts that density-dependent populations will have

Table 11.1   Scaling behaviour of the linear density-dependent model

Time scale (in terms of the natural time scale $1/r$		Power law for increments $\Delta x = x(t + \Delta t) - x(t)$
Short times	$(\Delta t \ll 1/r)$	$\Delta x \approx \text{const} \times \Delta t^{1/2}$
Intermediate times	$(\Delta t \approx 1/r)$	$\Delta x \approx \text{const} \times \Delta t^{H}$ $(0 < H < \frac{1}{2})$
Long times	$(\Delta t \gg 1/r)$	$\Delta x \approx \text{const}$ $(H = 0)$

$H < \frac{1}{2}$ over intermediate and long time scales. For time steps Δt very long compared with $1/r$, the values of $x(t)$ and $x(t + \Delta t)$ are asymptotically independent, and the scaling behaviour agrees with the general formula (11.2) for independent samples from a bounded random variable.

The model is in agreement with Sugihara and May's (1990a) thesis that if two species have different exponents H (and equal initial population sizes) then the one with the larger exponent H has greater likelihood of extinction. Moreover, the model gives an additional insight into this result. Given two species, the species with the stronger density dependence, and consequently the stronger return to equilibrium (larger value of r, or more negative eigenvalue $-r$), will have a smaller scaling exponent over any fixed time scale, although this effect will be very slight over very long time scales. In principle, the eigenvalue $-r$, which measures the Lyapunov stability, can be estimated by using regression to estimate the exponential decay of the correlation toward $-\frac{1}{2}$. Thus Sugihara and May are in effect measuring the strength of density dependence, and the stronger the density dependence, the slower will be the growth of variance.

11.3.1 Non-Brownian noise

A thought-experiment shows that similar results hold even in the presence of non-Brownian fractal noise with Hurst exponent H_0. Short-time $(\Delta t \ll 1/r)$ scaling behaviour must reflect only the scaling behaviour of the noise, and thus be characterized by the same Hurst exponent. Long-term behaviour $(\Delta t \gg 1/r)$ must follow the general formula (11.2), and yield H approaching zero. Finally, H must decay smoothly from its short-term value H_0 to its long-term value of 0 as the time step Δt is increased.

11.3.2 Nonlinear models

It is easy to see that a broad class of nonlinear models, including density-vague models, have the same three scaling regions as the linear density-dependent model wih additive noise. If the model under study has an equilibrium, and a characteristic time scale t_0 for the return to equilibrum

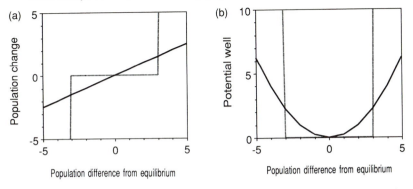

Fig. 11.1   Density-dependent and density-vague dynamics. (a) Density dependent (solid) and density vague (dashed) control. (b) Equivalent potential wells. Density-dependent dynamics is equivalent to a random walk in the solid potential well; density-vague dynamics to a random walk in the dashed potential well.

($t_0 = 1/r$ in the linear model), then fluctuations over very short times ($\Delta t \ll t_0$) reflect only the noise. Fluctuations over very long times ($\Delta t \gg t_0$) reflect the long-term variance of population, and there is an intermediate scaling region with intermediate behaviour. Consider, however, the density-vague model in which the population fluctuates randomly between a lower level x_L and an upper level x_H. Suppose that the noise and thus the short-term fluctuations are a random walk ($H = \frac{1}{2}$). Then the scaling property $x \approx t^{1/2}$ of random walks implies that the time τ required for the population level to reach one of the boundaries (x_L or x_H) is of order

$$\tau = (\tfrac{1}{2}(x_H - x_L))^2/\sigma_1^2. \tag{11.19}$$

For times $t < \tau$ the scaling behaviour of population fluctuations reflects that of the noise since there are no other expected effects, and the expected scaling exponent H is just $\frac{1}{2}$. For $t > \tau$ the boundaries play a role. In the case of reflecting boundaries, the population $x(t)$ at time t is a sample from a uniform distribution on the interval $x_L < x < x_H$. Figure 11.1a compares the behaviors of the density-dependent and density-vague models. The physical analogue of the linear density-dependent model consists of a ball undergoing a random walk in a parabolic potential well (see Fig. 11.1b). Similarly, the density-vague model corresponds to a random walk in a rectangular potential well.

11.4 Testing the models

We shall now test fractal and density-dependent models on real populations and simulated random walks and simulated density-dependent data. As described in the introduction to this chapter, we shall study 13 bird population time series from Holmes *et al.* (1986) and two butterfly time series from Harrison *et al.* (1991) and Ehrlich (personal communication). The bird populations are divided into two groups, one consisting of those species which went locally extinct (at least one census found no breeding pairs) during the study period and the other of those which did not go locally extinct. Some species which persisted throughout the study period might have gone extinct immediately after the study period. In addition, recall that we are seeking predictors of local extinction events. *We therefore did not use data from the last year of any population which persisted throughout the study or data from local extinction events themselves.* As is the case in all analysis of short time series, the results may be sensitive to our decision not to use these data.

The simulated data consisted of simulated random walks and simulated density-dependent models. Both models are subsumed under the formula

$$\Delta x = rx + \Delta w. \qquad (11.20)$$

In the random walk case, $r = 1$. In the density-dependent case, $0 < r < 1$ at short time scales, and r is negative at scales so long that the population overshoots the equilibrium. Most of the simulations use the random number generator in Turbo Pascal. An alternative random number generator (Wickman and Hill 1987) was used to test the dependence upon choice of random number generator in selected cases.

We seek to answer the following questions.

1. Which of the proposed models can be reasonably applied to the data?
2. Among the reasonable models, which best fits the data?
3. What does each of these models tell us about other models?
4. How can scaling exponents be best computed? What is the uncertainty?
5. What are the best predictors, if any, of population extinction?

Questions (1), (2), and (4) are important and naturally related. Many authors have simply computed a scaling relation over a large range of scales, and observed a good fit to a power law. These authors have simply ignored additional model verification. This works well if there is a good theoretical argument for a power law, and if there are many data points. However, here there are reasonable arguments for each of two competing models, and there are relatively few data points for each population. It is therefore especially important to carefully verify the hypotheses of any model.

As in Chapter 10, one must decide whether to do any preprocessing of the data. On the other hand, it also appears reasonable on general theoretical grounds to model population fluctuations multiplicatively, and to thus study the fluctuations of logarithms of population values, that is to 'log transform' all population data. It appears reasonable in terms of both resource availability and traditional density dependence to work directly with non-log-transformed data. We note that since the logarithm is approximately linear over a range of 3 or 4 to 1, it should make little difference whether the data are log-transformed or not. In order to check this, we begin by computing the Hurst exponent, using both raw and log-transformed data. (In fact, we found little difference between results obtained from raw and log-transformed data. Therefore, subsequent calculations used only raw data.)

11.5 Fractal versus density-dependent models; methodology

We use the following techniques to fit and test the applicability of fractal and linear density-dependent models.

11.5.1 *Computing the Hurst exponents*

We compute the Hurst exponents in all three ways described in Chapter 4 (see also Section 6.2): Hurst's growth of range technique, a related growth of moment technique, and a local technique which computes H in terms of the expectation of the product of successive increments. The first two compute the exponent H as an implicit 'average' over a fixed scaling region. The last computes a separate value of H for each time increment used.

Since the time series of bird populations covered a maximum of 16 years, we shall use a 5-year scaling region for the first two computations. Since Ehrlich (personal communication) observed that the computed exponent may depend upon the starting point, it is important to use all possible starting points within each time series for each time increment used. For example, consider a time series of 10 data points, one point for each year, after the last data point is deleted as explained above. For a 1-year time increment there are 9 possible starting points, namely $t = 1, 2, \ldots, 9$. For a 4-year increment there are only 6 possible starting points, namely $t = 1, 2, \ldots, 6$, since starting after year 6 leaves less than 4 years remaining. In general, for a time increment of length h, and time series of length T, there are $T - h$ possible starting points. For each species and time increment, we compute the average second moment over all possible starting points, and the average range (defined moment over all possible starting points, and the average range (defined as the difference between the maximum value and the minimum value of the population in the time increment under study) over all possible starting points. Under the fractal hypothesis, the second moment should scale as Δt^{2H}, and the range as Δt^H, where Δt is the time increment and H is the fractal exponent. The empirical exponent H is computed by log-transforming the time increments, second moments, and ranges to obtain models

$$\log(\text{second moment}) = \text{const} + 2H \log(\Delta t) \qquad (11.21)$$

and

$$\log(\text{range}) = \text{const} + H \log(\Delta t), \tag{11.22}$$

and fitting the above log-transformed models with linear regression. We do this for both raw and log-transformed population data (see also Section 4.2).

For short time series of real data, computation of the growth of range (formula (11.22)) appears to offer an advantage over computation of the growth of second moment (formula (11.21)) since the average range (as we have defined it) is automatically a nondecreasing function of the time lag for any time series. The application of formulae (11.21) and (11.22) in many cases yield different values of H. This suggests that the data are not fractal and also calls for additional analysis of the validity of formula (11.22) if only a few values of Δt are used in the calculation. Details appear in the following section.

For each species, and for time increments of one and two years, we also compute the correlation ρ between successive population increments, under the assumption that the expected population increment is 0. This assumption is very difficult to test for short time series; nonetheless the method computes a local fractal exponent (see Section 4.2). The correlation and Hurst exponent H are related by the formula

$$2^{2H} = 2 + 2\rho. \tag{11.23}$$

As above, we form averages over all possible starting points.

If the populations were fractal, all methods should yield roughly the same values of H, and in particular ρ and H would be independent of the time interval used. However, according to formula (11.6) above, for density-dependent populations, ρ and thus H would be decreasing functions of the time interval used.

Simulations and further analysis will be used to establish baselines and to see how well the above theoretical results hold for the short time series under study. We shall use the standard random number generator in Turbo Pascal as well as another random number generator (Wickman and Hill 1987) to generate the random walks used in these simulations. (There was no difference between the results with the two random number generators.) One cannot overemphasize the need to establish the scaling behaviour of neutral random models in applications such as the present one.

11.5.2 *Other tests*

In addition, we shall test whether the population increments $y(t + h) - y(t)$ depend upon the starting population values $y(t)$ for one- and two-year time increments h. In the case of fractal models, the increments are independent of the starting values. This follows from the axioms in Section

2.4. The increments *do* depend on the starting values in the case of density dependence. Simulation and theoretical analysis will be used to establish the expected behaviour under the null hypothesis of no density dependence.

In equilibrium density-dependent models, population levels tend to return to their equilibrium levels. This simple observation suggests an equally simple test for density dependence (Tanner 1966). Assuming that the mean value of a population is a reasonable estimate of its equilibrium value, does the population value tend to decrease when it is above the mean and increase when it is below the mean? We shall apply this test to both real and simulated data.

In addition, we shall attempt to fit real population data to the linear density dependent model of Section 11.2. More precisely, we shall use linear regression to fit the data to the following discrete-time version of that model:

$$y(t + 1) - y_{eq} = \exp(-r)[y(t + 1) - y_{eq}] + \Delta w. \qquad (11.24)$$

11.6 The Hurst exponent: results

We first describe the results of several computations of the fractal exponent H for real populations. Recall that bird populations were divided into two groups—those which persisted and those which went locally extinct.

Fractal exponents computed from the growth of second moment showed little difference between the two groups. Surprisingly, fractal exponents computed from the growth of range are *higher* in the persistent species than in the species which became extinct. Moreover, the two methods do not yield consistent values, which suggests a nonfractal model.

Note, however, in a curious apparent paradox, that the exponents computed from short samples of random walks show the same inconsistency, although they become asymptotically consistent for longer walks. These results therefore neither confirm nor rule out fractal models.

In contrast the local growth of moment method yields a fractal exponent which depends upon the sampling interval, and the exponent seen to decrease with the sampling interval. This behaviour suggests a density-dependent (non-fractal) model.

In conclusion, computations of the Hurst exponent for real population data display inconsistencies which suggest that the data are not fractal. However, short samples from the simulated random walks, the simplest fractal models, display many of the same inconsistencies. These results and their implications will now be discussed in more detail.

Table 11.2 summarizes the results of studying the growth of second moment and the growth of range, using raw and log-transformed data.

The two methods used to compute the Hurst exponent generally yield different answers, with the second moment growing more slowly than the

Table 11.2   Computation of *H* from growth of second moment and growth of range. The coefficient of correlation (of linear regression applied to log-transformed increments) is shown in parentheses whenever it is less than 0.99; unreported values are at least 0.99. In one case, that of the dark-eyed junco, marked with an asterisk, the growth of second moment gave a Hurst exponent outside the allowed range $0 < H < 1$

Species	Growth of second moment		Growth of range	
	Raw data	Transformed data	Raw data	Transformed data
(Birds which did not go extinct)				
Swainson's thrush	.40 (.98)	.42	.69	.66
Wood thrush	.36 (.97)	.38 (.97)	.74	.72
Ovenbird	.28 (.94)	.32 (.96)	.63	.68
American redstart	.26 (.95)	.16 (.95)	.68	.63
Red-eyed vireo	.18 (.79)	.18 (.79)	.60	.61
Veery	.03 (.19)	.10 (.54)	.64	.67
Hairy woodpecker	0 (0)	.01 (.09)	.34	.36
Median	.26	.18	.68	.67
(Birds which went extinct)				
Least flycatcher	.29 (.90)	.28	.72	.58
Downy woodpecker	.22 (.81)	.22 (.79)	.46	.51
Philadelphia vireo	.18 (.55)	.24 (.66)	.48	.48
Winter wren	.14 (.54)	.16 (.59)	.50	.47
Hermit thrush	.10 (.55)	.10 (.40)	.69	.69
Dark-eyed junco	.02 (−.08)*	−.02 (−.02)*	.54	.57
Median	.16	.19	.52	.54
(Butterflies—two colonies of bay checkerspots)				
JRC*	.09 (.87)	.32	.71	.62
JRH	.31	.24 (.98)	.70	.59

* See Harrison *et al.* (1991) for notation.

range. This suggests that the time series are not fractal. In order to find baselines for the above computations, we also apply the two methods of computing the fractal exponent to randomly generated time series of several lengths. The first set of simulations used the random number generator 'random' in Turbo Pascal. The results of these simulations are summarized in Table 11.3.

There appears to be a large difference between the computed and theoretical behaviours of the growth of range using a maximum lag of 5.

Table 11.3 Fractal exponents for simulated random walks of various lengths and maximum lags. Data are shown as median values followed by the maximum and minimum for ten replicates of each length and maximum lag. Taking longer series is similar to averaging over shorter series

Time series		Method for computing H	
Length	Maximum lag	Growth of second moment	Growth of range
16	5	.31 ($-$.07 to .76)	.66 (.38 to .94)
100	5	.54 (.09 to .63)	.73 (.59 to .80)
1000	5	.50 (.48 to .53)	.73 (.72 to .74)
1000	25	.50 (.46 to .54)	.66 (.64 to .67)

The expected value of the scaling exponent of a random walk is 0.5, and the observed values were typically between 0.63 and 0.75. The growth of second moment for the shortest series also appeared somewhat too slow (experimental $H = 0.31$ versus theoretical $H = 0.5$); this is not evident in the longer series. Therefore the growth of range calculations are repeated using a different random number generator (Wickman and Hill 1987). Similar results were obtained. This suggests that the second moment of increments of a short random walk grows more slowly than their range grows. We then analyse a discrete-time discrete-space random walk with time step 1 and spatial step ± 1 by enumerating all possible short sequences of spatial increments, and computing the expected range $R(\Delta t)$ for short time lags. The results presented in Table 11.4 are obtained.

Table 11.4 Growth of range $R(\Delta t)$ for Δt at most 5, compared with the value $(\Delta t)^{1/2}$ expected from the axioms for a continuous-time Brownian process

Time lag, Δt	Expected $R(\Delta t)$ from analysis of all paths	$(\Delta t)^{1/2}$
1	1.0	1.0
2	1.5	1.414
3	2.0	1.732
4	2.375	2.0
5	2.75	2.236

The expected value for the Hurst exponent associated with a short discrete-time discrete-space is readily obtained by fitting the data in the second column of Table 11.4 to the equation

$$R(\Delta t) = \text{const} \times (\Delta t)^{1/2}. \tag{11.25}$$

This yields $H = 0.63$ for the growth of range, in close agreement with the random walk simulations. The simulations and theory show the need for care in analysing the scaling behaviour of short time series. The significance of these results will be discussed below.

11.6.1 *Comparison of simulated and real data*

The two methods for computing the value of H show relatively large differences when applied to real data. In addition, the values of H obtained from the growth of second moment are generally less than $\frac{1}{2}$, the value expected for a random walk, and the values of H from the growth of range are generally more than $\frac{1}{2}$. This is most likely due to the short durations of samples and lags since similar effects were seen for short simulated random walks. As the length of simulations was increased, the range in observed values of H falls. This is an expected consequence of sampling, since the number of data points averaged in each computation of the range grows with increases in the duration of the time series under study. It is reasonable to expect that, for significantly longer simulations, the value of H computed from growth of range would approach the theoretical value of 0.5.

The observed values of H for real data are comparable to those for random walks. There appeared to be no significant differences between populations which went extinct and populations which did not go extinct. The power law for the growth of range consistently showed better fits than the power law for the growth of second moment. This suggests that the best approach to computing fractal exponents for short time series is to compute the growth of range. However, one must then use simulation or other techniques in order to relate the results obtained to those expected for various models.

We next compute fractal exponents with a local method: using the relative expectation of products of successive increments in the population value (see Section 4.3). This yields Hurst exponents H for each time increment, as presented in Table 11.5.

As above, we also repeated the calculations with data obtained from simulated random walks for comparison, obtaining roughly similar results.

However, for nine of the 15 populations (13 bird species plus 2 colonies of bay checker-spot butterflies) under study the local Hurst exponent H decreased as the lag was increased from one year to two years. In a random walk, there is no tendency for H to increase or decrease, and therefore the theoretical expected number of such decreases follows a binomial distribution

Table 11.5 Computation of H using the local growth of moment method for time increments (lags) of one and two years. Computations using the average growth of second moment from Table 11.3 are included for comparison. Note that ρ is not the coefficient of correlation unless the expectation of population increments is 0. In the axioms for fractals, formula (2.26) requires $H \geqslant 0$ and thus $\rho \geqslant -0.5$. Cases where this does not hold are indicated by asterisks

Species	Lag = 1		Lag = 2		H from average growth of moment
	ρ	H	ρ	H	
(Birds which did not go extinct)					
Swainson's thrush	−.08	.44	−.33	.21	.40
Wood thrush	−.18	.36	−.19	.35	.36
Ovenbird	−.28	.26	−.35	.19	.28
American redstart	−.21	.32	−.38	.14	.26
Red-eyed vireo	−.24	.30	−.38	.16	.18
Veery	−.22	.32	−.68	−.33*	.03
Hairy woodpecker	−.58	−.12*	−.29	.26	0
Median		.32		.26	.26
(Birds which went extinct)					
Least flycatcher	−.09	.43	−.25	.29	.29
Downy woodpecker	−.45	.07	.36	.72	.22
Philadelphia vireo	−.70	−.38*	−.21	.33	.18
Winter wren	−.60	−.15*	−.14	.39	.14
Hermit thrush	−.32	.21	−.70	−.26*	.03
Dark-eyed junco	−.49	.02	−.41	.01	.02
Median		.04		.31	.16
(Butterflies—two colonies of bay checkerspots)					
JRC*	−.36	.18	−.55	−.07*	
JRH	−.18	.35	−.32	.23	

* See Harrison *et al.* (1991) for notation

$B(15, 0.5)$ and is thus equal to 7.5 ± 1.9. This suggests that there may be density-dependent effects, although the significance level is low. ($p > 0.15$ from the statistics of the distribution).

11.7 Is there density dependence?

We therefore test for density dependence, testing both the applicability of the density-dependent model, and the fact that the increments of a fractal

Table 11.6 A study of the correlation between the population size and the population change over the next year. A change was classified as 'away from mean' if the population moved away from its mean value, and 'towards mean' if the population moved toward its mean value. Those cases in which there were more than twice as many moves towards the mean as moves away from the mean are flagged with asterisks

Species	Away from mean	No change	Towards mean
(Birds which did not go extinct)			
Swainson's thrush	4	1	9*
Wood thrush	2	2	10*
Ovenbird	1	4	9*
American redstart	5	1	8
Red-eyed vireo	4	0	10*
Veery	2	4	8*
Hairy woodpecker	5	4	5
Totals	23	16	59*
(Birds which went extinct)			
Least flycatcher	4	1	7
Downy woodpecker	2	5	4*
Philadelphia vireo	5	4	4
Winter wren	2	1	5*
Hermit thrush	5	1	3
Dark-eyed junco	0	1	6*
Totals	18	14	29
Bird grand totals	41	30	88*
(Butterflies—two colonies of bay checkerspots)			
JRC*	11	0	17
JRH	10	1	17

* See Harrison *et al.* (1991) for notation

process are independent of the values of the process itself. We first ask whether the population tends to decrease when it is above the mean and to increase when it is below the mean, a characteristic of all density-dependent models. The results are summarized in Table 11.6.

We next repeat the calculations for simulated data (see Table 11.7).

In the case of a random walk, one would expect the number of population

Table 11.7 A study of the correlation between the population size and the population change over the next year for simulated random walks and the density-dependent model (11.5): $dx = -rx\,dt + dB$. The value $r = 0$ is a continuous-time random walk (Brownian motion); if $r \neq 0$ the model has a natural time scale of $1/r$ years. Data are shown as totals from 30 simulations of each model; each simulation for 15 years (14 increments in population value)

Model	Away from mean	Towards mean
Random walk density-dependent, with natural time scale:	183 (43.6%)	237 (56.4%)
4 years ($r = 0.25$)	157 (37.4%)	263 (62.6%)
2 years ($r = 0.5$)	142 (33.8%)	278 (66.2%)
1 year ($r = 1$)	129 (30.7%)	291 (69.3%)

changes away from the mean to equal the number of population changes toward the mean. However, consider a case where the population has just jumped above the mean, then moves up once and down once to cross the mean and become less than it. There are two moves towards the mean and one move away from the mean. In general, each time the population crosses the mean there will be one more move towards the mean than away from it. In the case of long time series, there will be relatively large excursions and relatively few times when the population crosses the mean compared with the case of short time series. Thus the expected distribution of moves towards and away from the mean in the limit of long series need not hold for short time series. In fact, in a more extensive simulation of 10 000 short continuous-time random walks of 14 time steps, the population moved toward the mean 63.3% of the time. Table 11.8 provides additional examples.

The bird data showed no change in population 18.8% of the time (30 of 159 cases), perhaps because the populations were so small. We therefore performed a final simulation: 1000 integer-valued random walks, each involving 15 points or 14 changes, with the additional property that the population remained constant approximately 18–19% of the time. The actual results were as follows:

6616 steps toward the mean	(47.25%)
2547 steps with no change	(18.20%)
4837 steps away from the mean	(34.55%).

Table 11.8 The percentage of moves toward the mean for time series obtained from sampling continuous-time random walks as a function of the number of time steps n. Data are averages of 10 000 replicates and are presented in the form: means $\pm$ standard deviation

n	Percentage of moves towards mean
5	73.9 ± 15.5
10	65.2 ± 12.1
15	62.0 ± 10.3
25	60.3 ± 7.7
50	57.4 ± 5.6
100	55.2 ± 4.0
1000	51.7 ± 1.3

We first analysed the bird data using the χ^2 test with two degrees of freedom. First, there is a small but apparently not significant difference between the birds which persisted and those which went extinct ($\chi^2 = 0.96$, $p > 0.5$). Actual data are then compared with simulated data as a null hypothesis. A significant excess of moves toward the mean was interpreted as evidence of density dependence. The bird populations which persisted are density dependent ($\chi^2 = 7.19$, $p < 0.05$), whereas those which went extinct are less density dependent ($\chi^2 = 1.22$, $p > 0.4$, not significant). Overall the bird populations displayed a small amount of density dependence: there is a marginally significant excess of moves toward the mean ($\chi^2 = 6.14$, $p \approx 0.05$). Thus density dependence on a one-year scale may be correlated with persistence.

However, this distinction is lost when the population changes over two-year periods are similarly analysed: for both sets of bird population values moves toward the mean 66% of the time and away from the mean only 22% of the time. What is the significance of this apparent loss of density dependence?

We next compute the Lyapunov exponent $(-r)$ assuming a linear density-dependent model, both to determine an appropriate time scale and to learn more about density dependence. More precisely, we fit the data to a discrete-time version of the linear density-dependent model

$$x(t + 1) = r'x(t) + \Delta w, \tag{11.26a}$$

where, as usual, x denotes the difference between a population level and its 'equilibrium' value, and

$$r' = \exp(-r). \qquad (11.26b)$$

The average population level is used as an estimate of the equilibrium. Note that the parameter r' above *may be negative*, indicating that one-year population changes overshoot the mean: with the population changing from above the mean to below the mean or vice versa. In these cases the implicit one-year time scale in equation (11.26b) is too long to reliably apply linear models and thus determine the Lyapunov exponent. Otherwise, r' above is positive and is related to the Lyapunov exponent in the original continuous-time model (11.4) by the formula

$$r' = \exp(-r), \quad \text{or equivalently } r = -\ln(r'). \qquad (11.27)$$

For comparison, we also use a two-year time delay, and compare the resulting parameter r' with its expected value in the absence of significant effects due to time delays, namely, the square of the value of r' in the model with a one-year delay (11.26b). The results are presented in Table 11.9.

Again, real data appeared significantly different from simulated random walks. The natural time scale of the discrete time model is given by the formula

$$r'^{t_0} = 1/e, \quad \text{or equivalently } t_0 = -1/\ln r'; \qquad (11.28)$$

compare formula (11.6). The apparent natural time scale of bird populations which went extinct was about $\frac{1}{2}$ year, half the apparent time scale (0.93 year) of bird populations which did not go extinct. We use the phrase 'apparent time scale' because both time series represent only annual data, making it difficult to study systems with faster dynamics.

Moreover, the eigenvalue r' associated with a two-year unit time step in a *linear* difference equation model should be the square of the eigenvalue associated with a one-year unit time step since a two-year time step corresponds to two successive one-year steps. An informal examination of the data in Table 11.9 shows significant deviations from this behaviour in all 6 bird populations which went extinct, but only 2 of the 7 bird populations which did not go extinct, and in neither butterfly population. These results suggest that the combination of nonlinearities with short time scales may make extinction more likely. May's (1974) work on ecosystem dynamics suggested that the combination of nonlinearities and significant time delays might yield chaotic dynamics, reminiscent of the discrete-time logistic equation (cf. Collet and Eckmann 1980).

Table 11.9 The parameter r' in the *discrete-time* linear model $x(t + \mathrm{lag}) = r'x(t) + \Delta w$, for lags of one and two years. Apparently significant anomalies in the two-year value are denoted by asterisks

Species	Lag		
	1 year, r'	1 year, r'^2	2 years, r'
(Birds which did not go extinct)			
Swainson's thrush	.85	.73	.79
Wood thrush	.63	.39	.34
Ovenbird	.41	.16	.11
American redstart	.32	.10	.10
Red-eyed vireo	.34	.11	.020
Veery	.086	.0073	−.42*
Hairy woodpecker	0	0	.20*
Median	.34	.11	.10
(Birds which went extinct)			
Least flycatcher	.75	.56	.39*
Downy woodpecker	0	0	.17*
Philadelphia vireo	.48	.22	.67*
Winter wren	−.08	.0064	.38*
Hermit thrush	.19	.038	−.17*
Dark-eyed junco	−.27	.073	−.11*
Median	.06	.0036	.25
(Butterflies—two colonies of bay checkerspots)			
JRC*	.21	.044	−0.023
JRH	.60	.35	.34
(Random walk simulations of various lengths)			
10 years	.47 ± .10	.22 ± .10	.15 ± .15
15 years	.66 ± .073	.43 ± .10	.41 ± .16
100 years	.94 ± .002	.88 ± .004	.88 ± .006

* See Harrison *et al.* (1991) for notation

11.7.1 *The time scale of density dependence.*

Although bird populations which persisted throughout the study appeared to show apparently stronger density dependence over one-year periods than those which went locally extinct, this difference was lost over two-year periods. We now relate this behaviour to the natural time scales of the populations obtained using the linearized analysis of formulae (11.26) through (11.28).

Both sets of populations had time scales of one year or less. These time scales may be too short for samples at two-year periods to uncover significant differences in 'linear' features in the dynamics such as density dependence (see Figure 11.1). This effect is especially pronounced in the presence of significant non-linear effects shown in Table 11.9.

In addition, possible chaotic dynamics may obscure density dependence over longer time scales, since the dynamics of a chaotic system depends sensitively upon the initial conditions. Thus, especially in the presence of noise, the linkage between the population value at time t and its value at time $t + 1$ may be lost at time $t + 2$, making it hard to see density dependence over a two-year scale.

11.8 Discussion

We examined several models for population fluctuations: the fractal model introduced by Sugihara and May (1990a), a linear density-dependent model, and general nonlinear models. Our time series are too short for nonlinear techniques, but they may be the most promising for other applications.

The scaling behaviour of real populations appeared more consistent with the density-dependent model than with the fractal model. Moreover, by one simple test for density dependence, namely the tendency to move toward the mean, the degree of density dependence over one-year time lags appears to correlate with persistence. Unfortunately, this distinction is lost over two-year periods, perhaps as a result of sampling problems or underlying chaotic dynamics.

Persistent populations also displayed longer time scales and a significantly closer fit to linearity than species which went extinct. Perhaps extinction arises from a combination of nonlinearities and time scales much shorter than the annual cycle of temperate woodland birds, a combination which can yield chaotic dynamics. It is interesting that the apparent time scale of persistent species was approximately one year, as might be expected from the biology.

The fractal exponent remains a potentially useful tool even in the case of density dependence, since computation of the fractal exponent can locate scaling regions and help estimate the long-term range of population fluctuations.

11.9 Appendix

Table 11.10 lists the bird population data, from Holmes *et al.* (1986), suggested by Ehrlich (personal communication), used in this study. Each time series consists of annual data, starting in 1969. We invite readers to try other techniques for analysing these data.

Table 11.10   Bird population data, following Holmes *et al.* (1986)

Species	Population time series (truncated at extinction)
(Birds which did not go extinct)	
Swainson's thrush	9, 8, 11, 11, 9, 7, 8, 7, 4, 5, 7, 3, 2, 1, 2, 1
Wood thrush	7, 6, 9, 7, 4, 5, 6, 5, 9, 7, 7, 3, 1, 2, 2, 1
Ovenbird	6, 6, 9, 9, 11, 10, 16, 16, 9, 13, 12, 10, 10, 11, 14, 11
American redstart	12, 26, 29, 29, 26, 22, 39, 42, 44, 32, 36, 34, 22, 32, 30, 14
Red-eyed vireo	20, 24, 29, 22, 26, 23, 31, 30, 24, 20, 16, 13, 26, 23, 22, 21
Veery	2, 2, 5, 2, 1, 1, 3, 3, 3, 4, 3, 1, 2, 3, 2, 2
Hairy woodpecker	2, 1, 1, 2, 2, 1, 2, 2, 3, 2, 3, 2, 2, 3, 1, 2
(Birds which went extinct)	
Least flycatcher	26, 28, 43, 50, 56, 26, 30, 28, 34, 22, 22, 15, 2, 0
Downy woodpecker	6, 3, 3, 2, 2, 2, 3, 3, 3, 4, 2, 4, 0
Philadelphia vireo	4, 7, 5, 8, 4, 4, 4, 4, 2, 3, 2, 3, 2, 2, 0
Winter wren	1, 1, 2, 1, 3, 6, 2, 5, 1, 0
Hermit thrush	6, 8, 7, 7, 4, 2, 8, 5, 4, 2, 0
Dark-eyed junco	8, 4, 8, 8, 2, 5, 5, 6, 2, 0

Part V

The toolbox

We conclude with a toolbox for fractal modellers: Chapter 12 contains documented listings and descriptions of many of the programs used above. Many books would call Chapter 12 an appendix; we include it in the body of the text to emphasize both the ease of fractal modelling and the role of hands-on experiments for students and practitioners. The toolbox is thus both the end of the book and the beginning of fractal analysis of your data.

12

Programs

12.1 Introduction

We conclude this book on 'fractals for users' with illustrative ready-to-use programs for the construction of regular and random fractals, and fractal analysis of patterns and time series. These programs were used in the previous chapters, and are written in Turbo Pascal to make them readily accessible to all.

There are several programming environments in common use, including UNIX and three microcomputer environments: Microsoft DOS on Intel-based machines such as the IBM PC and clones, Apple's operating system on MacIntosh computers, and most recently Windows on Intel-based machines. The DOS interface uses text and commands similar to those on most mainframe and UNIX terminals. Apple's system and Windows use 'graphical user interfaces'; the user uses a mouse to select and enter commands on the screen. The latter two environments offer generally more user-friendly applications software, at a cost in programming complexity. Much of the programming effort for MacIntosh and Windows environments goes to the interface between the user and the software—input and output. It appears convenient to us to describe algorithms with Turbo Pascal programs running under the Turbo editor or DOS. They are readable by those with little or no experience and can be translated to more efficient (but harder to read) languages such as C by the experienced user. Moreover, although the programming environments are not compatible at the machine code level, our Pascal code will run in most environments with little or no modification except in graphics instructions.

This chapter is organized as follows. Section 12.2 presents a detailed development of a program for the Koch snowflake in order to demonstrate one simple way to implement iteration. A program for the randomized Koch snowflake of Section 2.4 is included for comparison. In order to keep this chapter relatively short, subsequent sections are less detailed. Section 12.3 contains programs for two cellular automata, the one-dimensional game of life, one of the simplest examples (see Section 7.2) and the Bak–Tang–Weisenfeld sandbox model (see Section 8.2). Random fractals are illustrated with a program for random walks in Section 12.4 and for Mandelbrot–Weierstrass fractals in Section 12.5. Section 12.6 contains programs for the fractal analysis of patterns, beginning with a program to fit hyperbolic distributions with log transforms and linear regression (see Section 2.6 and Chapter 6). Section 12.7 contains programs for the fractal analysis of time series. A few additional programs are included in Section 12.8.

12.2 Recursion and iteration

The Koch snowflake is typical of the fractals constructed through recursion or iteration. The construction also illustrates both scale-invariance and the sharp differences between fractal curves and the smooth and polygonal curves of Euclidean geometry. The 'classic' Koch snowflake is a limit in an appropriate sense of an iterative process which begins with an equilateral triangle. Each iteration is constructed by replacing each side of the previous iteration by a four-sided polygonal path called the generator. Each side of the generator is one-third as long as the distance between the endpoints of the generator, or equivalently between the endpoints of the line it replaces. The generator has three turns, which we regard as 'left 60°', then 'right 120°', and finally 'left 60°' again, and label simply as 'L, R, L' (see Fig. 12.1). In this language, the first stage of the Koch snowflake, an equilateral triangle, is represented by the sequence R, R, R. To draw this stage, choose a unit length s, a starting point, and a starting direction. Then, beginning at the starting point, perform the following process three times:

> begin
> > draw a line segment of length s;
> > turn right 120°;
> end.   (12.1)

To construct the second stage, insert the sequence of turns 'L, R, L' before each 'R' in the original sequence, and multiply the starting length by $\frac{1}{3}$. This yields a new sequence,

$$L, R, L, \underline{R}, L, R, L, \underline{R}, L, R, L, \underline{R}, \qquad (12.2)$$

in which the original sequence of three right turns is underlined. This stage may be drawn by reading the sequence (12.2) and performing the following process 12 times:

> begin
> > draw a line segment of length s;
> > turn right 120° or left 60° as indicated in sequence (12.2);
> end.   (12.3)

The above ideas are easily turned into a program for drawing a Koch snowflake. Our program uses iteration rather than recursion to make it more accessible to the reader with little programming experience.

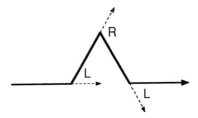

Fig. 12.1   The generator of the Koch snowflake. In traversing the generator from left to right, one makes 60° left turns at the points labelled L, and a 120° right turn at the point labelled R.

12.2.1 *Program koch*

```
program koch;
      (* draws 4 stages in the Koch snowflake *)

uses
      graph3, crt;
      (* Turbo Pascal 5 command to include appropriate
      modules in compiled program.  The module graph 3 lets
      us use the more intutitive graphics commands from
      Turbo Pascal 3. *)

var
      turns, oldturns : array [1..192] of char;
                              (* store the sequences of turns
                              described above *)
      stage : integer;
      numturns : integer;    (* the number of turns in the
                              current stage *)
      s : integer;           (* the length of an edge in the
                              current stage *)
      i,j : integer;
      direction : real;      (* the current direction in
                              degrees counterclockwise from
                              the positive sense of the
                              x-axis *)
      u,v,uold,vold : real; (* coordinates for graphing *)

procedure drawstage;
      (* draws one stage of the snowflake *)

      begin
            graphcolormode;     (* initializes graphics mode and
                              clears the screen *)
```

```
(* now describe the starting point and direction *)
uold := 160;
vold := 195;
direction := 4*pi/3;

(* construct and plot lines *)
for i := 1 to numturns do      (* i   denotes the
                    number of the line *)
   begin
      u := uold + s * cos(direction);
      v := vold + s * sin(direction);
      draw(round(uold),round(vold),round(u),
            round(v),3);    (* line from (uold,vold) to
                       (u,v) in color number 3 *)

            (* now prepare for next line *)
            uold := u;
            vold := v;
            if turns[i] = 'L'
               then direction := direction - pi/3
                     (* turn left 60 deg *)
               else direction := direction + 2*pi/3;
                     (* turn rt 120 deg *)
         end;   (* for i := 1 to numturns *)
   end;   (* procedure drawstage *)

begin (* main program *)

   writeln('Program will draw 4 stages of snowlake.  Type
      "enter" to');
   writeln('clear screen after each stage');
   writeln('Type "enter" to begin.');
   readln;

   (* first describe stage 0 *)
   numturns := 3;
   for i := 1 to numturns do
      turns[i] := 'R';
   s := 162;             (* chosen so that s/3, s/9, and s/27
                     will be integers *)

   drawstage;
   readln;             (* wait for user to type "enter" to
                     continue *)

   for stage := 1 to 3 do
      begin
         for i := 1 to numturns do
            oldturns[i] := turns[i];  (* save list of
                     turns *)
         (* now construct list of current turns, as
               described above *)
```

```
        j := 1;
        for i := 1 to numturns do
            begin  (* insert L, then R, and then L
                    before oldturns[i] *)
                turns[j] := 'L';
                j := j + 1;
                turns[j] := 'R';
                j := j + 1;
                turns[j] := 'L';
                j := j + 1;
                turns[j] := oldturns[i];
                j := j+1;
            end;  (* for i := 1 to numturns *)

        (* rescale *)
        numturns := 4 * numturns;
        s := s div 3;
        drawstage;
        readln;      (* wait for user input to continue *)

    end;  (* for stage := 1 to 3 *)
    textmode(80);
end.  (* main program *)
```

A Koch snowflake is shown in Fig. 2.1 in Chapter 2.

12.2.2 *Variations*

(a) Both the initial stage and the generator can be varied.
(b) One can construct a randomized Koch snowflake as follows. Use a line
segment for the initial stage. Define a second generator 'R0, L0, R0'
which is the mirror image of the first generator by letting R0 denote a
60° right turn and L0 a 120° left turn. This generator is the mirror image
of the first generator. The above code is readily modified to draw such
a randomized Koch snowflake. Figure 2.5 shows a randomized Koch
snowflake.

12.2.3 *Program randomkoch*

```
program randomkoch;
    (* draws 4 stages in a randomized Koch snowflake *)

uses
    graph3, crt;

var
    turns, oldturns : array [1..192] of string[2];
                        (* store the sequences of turns
```

```
                             described above *)
    stage : integer;
    numturns : integer;    (* the number of turns in the
                              current stage *)
    s : integer;           (* the length of an edge in the
                              current stage *)
    i,j : integer;
    direction : real;      (* the current direction in
                              degrees counterclockwise from
                              the positive sense of the
                              x-axis *)
    u,v,uold,vold : real;  (* coordinates for graphing *)
    temp : real;           (* temporary *)

procedure drawstage;
    (* draws one stage of the snowflake *)

  begin
    graphcolormode;        (* initializes graphics mode and
                              clears the screen *)

    (* now describe the starting point and direction *)
    uold := 160;
    vold := 195;
    direction := 4*pi/3;

    (* construct and plot lines *)
    for i := 1 to numturns do     (* i denotes the
                          number of the line *)
       begin
         u := uold + s * cos(direction);
         v := vold + s * sin(direction);
         draw(round(uold),round(vold),round(u),
             round(v),3);   (* line from (uold,vold) to
                           (u,v) in color number 3 *)
       (* now prepare for next line *)
       uold := u;
       vold := v;
       if turns[i] = 'L'
          then direction := direction - pi/3;
               (* turn left 60 deg *)
       if turns[i] = 'R'
          then direction := direction + 2*pi/3;
               (* turn rt 120 deg  *)
       if turns[i] = 'RO'
          then direction := direction + pi/3;
               (* turn rt 60 deg *)
       if turns[i] = 'LO'
          then direction := direction - 2*pi/3;
               (* turn left 120 deg *)
    end;  (* for i := 1 to numturns *)
end;  (* procedure drawstage *)
```

```
begin (* main program *)

    writeln('Program will draw 4 stages of random
        snowflake.  Type');
    writeln('"enter" to clear screen after each stage.');
    writeln('Enter an integer (0 to 1000) to set random
        number generator.');
    readln(j);
    for i := 1 to j do
        temp := random;
    writeln('Type "enter" to begin.');
    readln;
    (* first describe stage 0 *)
    numturns := 3;
    for i := 1 to numturns do
        turns[i] := 'R';
    s := 162;                   (* chosen so that s/3, s/9, s/27
                                will be integers *)
    drawstage;
    readln;                     (* wait for user to type "enter"
                                to continue *)

    for stage := 1 to 3 do
        begin
            for i := 1 to numturns do
                oldturns[i] := turns[i];  (* save list of
                                    turns *)
            (* now construct list of current turns, as
                    described above *)
            j := 1;
            for i := 1 to numturns do
                if random < 0.5 then    (* use the original
                                    generator with probability
                                    0.5 *)
                    begin           (* insert L, then R, and then L
                                    before oldturns[i] *)
                        turns[j] := 'L';
                        j := j + 1;
                        turns[j] := 'R';
                        j := j + 1;
                        turns[j] := 'L';
                        j := j + 1;
                        turns[j] := oldturns[i];
                        j := j + 1;
                    end  (* for i := 1 to numturns *)

                else  (* use the mirror-image generator *)
                    begin           (* insert R0, then L0, and then
                                    R0 before oldturns[i] *)
                        turns[j] := 'R0';
                        j := j + 1;
```

```
                turns[j] := 'L0';
                j := j + 1;
                turns[j] := 'R0';
                j := j + 1;
                turns[j] := oldturns[i];
                j := j + 1;
            end;   (* for i := 1 to numturns *)

        (* rescale *)
        numturns := 4 * numturns;
        s := s div 3;
        drawstage;
        readln;      (* wait for user input to continue *)
    end;   (* for stage := 1 to 3 *)

    textmode(80);

end.  (* main program *)
```

12.3 Evolution: cellular automata

The next two programs illustrate the one-dimensional game of life of Section 7.2 and the sandbox model described in Section 8.2.

12.3.1 *Program for game of life (cellauto)*

```
program cellauto;
    (* Runs the one-dimensional game of life, a simple,
    totalistic cellular automaton. *)

uses
    crt, graph3;

var
    i,j,k,l : integer;
    state, newstate : array[0..319] of integer; (* state[i]
        is the present state of the CA at position  x = i,
        newstate[i] is the next state at position i, all
        states are updated in parallel *)
    init : char;       (* selects random or deterministic
                        initial conditions *)
    sum : integer;

begin
    (* initialize *)
    writeln('Enter R to start in a random initial state;
        otherwise enter D.');
```

```
readln(init);
writeln('At conclusion, type "enter" to clear
     screen.');
writeln('Type "enter" to begin.');
readln;           (* begin after "enter" *)

graphcolormode; (* sets screen to graphics mode *)
for i := 0 to 319 do (* initialize state arrays *)
    begin               (* all zero except for some *)
         state[i] := 0;       (* positions below *)
         newstate[i] := 0;
    end;
if ((init = 'R') or (init = 'r')) then  (* random
                    initial state *)
    begin
         for i := 0 to 319 do
             if random < 0.25 then state[i] := 1;
    end
    else
         state[160] := 1;
(* now run and display 200 time steps *)
for i := 1 to 318 do
    plot(i,0,state[i]);   (* see note below on
                        boundary conditions *)
for j := 1 to 199 do (* time j runs from 1 to 199 *)
    begin
         for i := 1 to 318 do    (* uses toroidal
                    boundary conditions, so that
                    position 0 is identified with
                    position 318 and position 1 is
                    identified with position 319 *)
             begin   (* apply the CA rule, see
                    Section 7.1 *)
                 sum := state[i-1] + state[i] +
                 state[i+1];
                 if ((sum = 2) or (sum = 1)) then
                    newstate[i] := 1
                 else newstate[i] := 0;
                 plot(i,j,newstate[i]);
             end;
         for i := 1 to 318 do    (* copy newstate
                    array to state array *)
             state[i] := newstate[i];
         state[0] := newstate[318];   (* applying
                    toroidal boundary conditions*)
         state[319] := newstate[1];
    end;
readln;                        (* to clear screen *)
textmode(3);
end.
```

12.3.2 *Program fragment for sandbox*

```
program sandbox;
      (* reader should add appropriate output--to screen or file *)

uses
      crt;

var
      i, j, k, l: integer;
      size: integer;
      z: array[1..64, 1..64]of integer;  (*array of states*)
      dz: array[0..65, 0..65] of integer;  (*array of state changes,
                                            includes border on array
                                            of states*)
      searchsize: integer;
      maxtime: integer;
      dropflag: integer;

procedure redistribute;  (*sand after energy gets too big*)
begin
      dz[j, k] := -4;
      dz[j+1, k] := 1;
      dz[j-1, k] := 1;
      dz[j, k+1] := 1;
      dz[j, k-1] := 1;
end;  (*procedure redistribute*)
begin (*main*)
      writeln('input size');
      readln(size);
      writeln('input the amount of steps');
      readln(maxtime);
      for i := 1 to size do
            for j := 1 to size do
                  z[i, j] := 0;
      for i := 0 to size+1 do
            for j := 0 to size+1 do
                  dz[i, j] := 0;
      for i := 1 to maxtime do
            begin
            (*drop sand*)
            j := trunc(size*random)+1;
            k := trunc(size*random)+1;
            z[j, k] := z[j, k]+1;
            ('does it collapse')
            if z[j, k] > 3 then
                  begin
                  redistribute;
                  (*follow cascade of sand*)
                  dropflag := 1:  (*sand dropped*)
                  repeat
                     dropflag := 0;  (*reset flag*)
                     for j := 1 to size do
```

```
            for k := 1 to size do
              z[j,k] := z[j,k]+dz[j,k];
          for j := 0 to size+1 do
            for k := 0 to size+1 do
              dz[j,k] := 0;
          for j := 1 to size do
            for k := 1 to size do
              if z[j,k]>3 then
                begin
                  dropflag := 1;
                  redistribute;
                end; (*if z[j,k]>3*)
        until dropflag := 0;

    end; (*for i*)

  (*insert output*)

end.
```

12.4 Random walks

We illustrate random walks and their diffusion limit, Brownian motion, the prototype fractal curve. The notation and mathematics follow that of Section 2.4.

12.4.1 *Random walk program (randomwalk)*

```
program randomwalk;
  (* graphs 3 random walks *)

uses
  crt, graph3;

var
  seed : integer;              (* seed for random number
                                  generator *)
  deltat, deltay : integer;    (* time and space steps *)
  t,y,oldt,oldy : integer;     (* coordinates of point on
                                  random walk in time
                                  and space (real line)  *)
  steps : integer;             (* number of steps in
                                  random walk *)
  i,j : integer;
begin
```

```
(* initialize *)
writeln('Program will draw 3 random walks.  After
  completion,');
writeln('type "enter" to clear screen.');
writeln('Input an integer (1 to 32767) to set the');
writeln('random number generator');
readln(seed);
steps := 16;
deltat := 16;
deltay := 8;


graphcolormode;
for i := 1 to 3 do  (* draw 3 random walks *)
   begin   (* construct this random walk *)
      randseed := seed;  (* so that each walk begins
                          with a rescaled version of the
                          previous walk *)
      oldt := 32;
      oldy := 100;
      for j := 1 to steps do
         begin
            t := oldt + deltat;
            if random < 0.5
               then y := oldy - deltay    (* down or
                        up, each with probability 1/2 *)
               else y := oldy + deltay;
            draw(oldt,oldy,t,y,i); (* draw the line
                        from (oldt,oldy) to (t,y) in
                        color number  i *)
               oldt := t;
               oldy := y;
            end;

         (* rescale for next random walk, illustrating
         diffusion limit *)
         deltay := deltay div 2;
         deltat := deltat div 4;  (* thus deltay**2/deltat
                                   = const *)
         steps := 4 * steps;
      end;  (* for i := 1 to 3 *)
   readln;  (* wait for user input to clear screen *)
   textmode(80);
end.
```

12.5 Mandelbrot–Weierstrass fractals

Mandelbrot–Weierstrass fractals (Berry and Lewis 1980), described in Section 5.6, provide the fastest implementation of random fractal curves and surfaces. Programs for generating randomly selected approximations to Mandelbrot–Weierstrass fractals with specified Hurst exponents follow.

12.5.1 *Program for Mandelbrot–Weierstrass fractals using the discrete Fourier transform (mwfractal).*

```
program mwfractal;
        (* graphs Mandelbrot-Weierstrass fractals in one
        dimension --   demonstration version using discrete
        Fourier transform, not fast Fourier transform *)

uses
    crt,graph3;

var
    y : array [0..319] of real;  (* coordinate on screen *)
    i,j : integer;
    h : real              (* Hurst exponent *)
    hplus : real;         (* Hurst exponent plus 1/2 *)
    terms : integer;      (* highest frequency in
                          spectrum, at most 256 *)
    a,p : array[1..256] of real; (* amplitudes and phases
                          of spectrum *)
    ymax : real;          (* maximum y to scale image *)
    scale : real;         (* to scale image *)
    temp : real;          (* temporary real *)
    twopidiv : real;      (* 2*pi / 320, so as to graph
                          one complete cycle in the 320
                          pixel width of the screen *)

begin
    (* initialize *)
    twopidiv := 2*pi/320;   (* Turbo Pascal has pi as a
        predefined const. *)
    writeln('What is the Hurst exponent H (0 to 1 (rough to
        smooth)) ?');
    readln(h);
    hplus := h + 0.5; (* to simplify calculating the
        spectrum *)
    writeln('Highest frequency (at most 256) ?');
    writeln('(Program will run slowly if the highest
        frequency exceeds 16.');
    writeln('The fft version    fftmwfractal    is
        faster.)');
    readln(terms);
    writeln('Input an integer (0 to 1000) to set random
        number generator.');
    readln(j);
```

```
for i := 1 to j do
    temp := random;
writeln('Program will draw fractal.  Type "enter" to
    clear');
writeln('screen and exit when done.');


(* generate the spectrum *)
for i := 1 to terms do
    begin
        temp := 0;
            for j := 1 to 12 do      (* generate sample from
                                      normal distribution N(0,1),
                                      as sum of 12 independent
                                      samples from uniform
                                      distributions *)
                temp := temp + random - 0.5;

            (* make amplitude of ith term of spectrum
            normal with mean 0 and variance 1/(i**h)
            and make phase random in units of pixels *)
            a[i] := temp*exp(-hplus*ln(i));
            p[i] := 320*random;
    end;

(* generate y-values by applying inverse Fourier
    transform *)
ymax := 0;      (* will hold maximum absolute value of y
    for scaling *)
for i := 0 to 319 do(* loop over x-values, in pixels *)
    begin
        y[i] := 0;
        for j := 1 to terms do
            y[i] := y[i]+ a[j]*cos(twopidiv*j*(i-p[j]));
        if abs(y[i]) > ymax then ymax := abs(y[i]);
    end;

for i := 0 to 319 do        (* scale y to the interval
                            [-90,90] *)
    y[i] := 90*y[i]/ymax;

(* plot the fractal *)
graphcolormode;
for i := 0 to 318 do
    draw(i,100-round(y[i]),i+1,100-round(y[i+1]),3);
        (* line from ith to (i+1)th point, vertical
        midpoint is y = 100, coordinates increase down
        the screen *)
readln;
textmode(80);
end.
```

REMARKS 12.1 The sum of twelve uniform random variables is approximately normal by the central limit theorem. If the uniform random variables take values in the interval $[-\frac{1}{2}, \frac{1}{2}]$, then the resulting normal distribution has mean 0 and variance 1.

The use of the fast Fourier transform yields a computationally faster, but more complicated, implementation of the program for Mandelbrot–Weierstrass fractals. In addition, this program can be used to generate fractal time series with known Hurst exponents for testing programs for the fractal analysis of time series (Section 12.6). Here is the code.

12.5.2 Program for Mandelbrot–Weierstrass fractals using the fast Fourier transform (fftmwfractal)

```
program fftmwfract;
      (* graphs Mandelbrot-Weierstrass fractals in one
      dimension using (inverse) fast Fourier transform *)

uses
    crt,graph3;

var
    y : array [0..256] of real; (* coordinate on screen *)
    i,j : integer;
    h : real;              (* Hurst exponent *)
    hplus : real;          (* Hurst exponent plus 1/2 *)
    terms : integer;       (* highest frequency in
                           spectrum, at most 127 *)
    areal,aimag : array[1..256] of real; (* real and imag
                           parts of spectrum *)
    ymax : real;           (* maximum y to scale image *)
    scale : real;          (* to scale image *)
    temp : real;           (* temporary real *)
    twopidiv : real;       (* 2*pi / 256, so as to graph
                           one complete cycle in the 256
                           pixel width used *)

    index : array [1..256] of integer;   (* for bit reversal
                           in fft *)
    costable, sintable : array [0..256] of real;   (* for
                           fft *)

    outfile : text;        (* for optional output file *)
    filename : string[80];
    flag : string[2];
```

```
const
   n = 256;              (* number of terms, size of fft *)
   hn = 128;             (* half n *)
   log2n = 8;            (* log to base 2 of n *)


procedure initbitrev; (* constructs index array for bit
      reversal in fft *)
   var k,l,lmax,ii,m : integer;

   begin
      index[1] := 1;
      lmax := 1;
      ii := hn;

      for k := 1 to log2n do
         begin
            for l := 1 to lmax do
               index[l+lmax] := index[l] + ii;
               ii := ii div 2;
               lmax := 2 * lmax;
            end;
   end;   (* procedure initbitrev *)

procedure trigtable;
   var
      k : integer;
      angle : real;

   begin
      for k := 0 to n do
         begin
            angle := 2*pi*k/n;
            costable[k] := cos(angle);
            sintable[k] := sin(angle);
         end;
   end;   (* procedure trigtable *)

procedure transform;
   var
      i,i0,i1,j,l,p,hp, ndivp : integer;
      tempreal, tempimag : array[1..256] of real;
      c,s : real;    (* cosine, sine *)
      t1, t2 : real;   (* temporaries *)

   begin
      (* first do bit reversal *)
      for i := 1 to n do
         begin
            i0 := index[i];
            tempreal[i] := areal[i0];
```

```
      tempimag[i] := aimag[i0];
   end;
for i := 1 to n do
   begin
      areal[i] := tempreal[i];
      aimag[i] := tempimag[i];
   end;
      (* initialize some variables *)
      p := 2;
      hp := 1;
      ndivp := n div p;

      for l := 1 to log2n do
         begin
            for j := 1 to hp do
               begin
                  c := costable[(j-1)*ndivp];
                  s := -sintable[(j-1)*ndivp];
                     (* inverse FFT *)
                  i := j;
                  for i1 := 1 to ndivp do
                     begin
                        i0 := i + hp;
                        t1 := areal[i0]*c - aimag[i0]*s;
                        t2 := areal[i0]*s + aimag[i0]*c;
                        areal[i0] := areal[i] - t1;
                        aimag[i0] := aimag[i] - t2;
                        areal[i] := areal[i] + t1;
                        aimag[i] := aimag[i] + t2;
                        i := i + p;
                     end;   (* for i1 *)
               end;   (* for j *)
            hp := p;
            p := 2*p;
            ndivp := ndivp div 2;
            end;   (* for l *)

end;   (* procedure transform *)

begin   (* main *)


   (* initialize *)
   twopidiv := 2*pi/256;   (* Turbo Pascal has pi as a
     predefined const. *)
   writeln('Enter the Hurst exponent H (0 to 1 (rough to
     smooth)).');
   readln(h);
   hplus := h + 0.5;   (* to simplify calculating the
     spectrum *)
   writeln('Highest frequency (at most 127) ?');
```

```
readln(terms);
writeln('Input an integer (0 to 1000) to set random
   number generator.');
readln(j);
for i := 1 to j do
   temp := random;
writeln('Program will draw fractal.  After completion,
   type "enter"');
writeln('to clear screen, and optionally write y-values
   to file.');

(* setups for FFT *)
initbitrev;
trigtable;

for i := 1 to n do
   begin
      areal[i] := 0;
      aimag[i] := 0;
   end;  (* for i *)

(* generate the spectrum *)
for i := 2 to terms+1 do
   begin
      temp := 0;
      for j := 1 to 12 do
         temp := temp + random - 0.5;
      areal[i] := temp*exp(-hplus*ln(i-1));
      temp := 0;
      for j := 1 to 12 do
         temp := temp + random - 0.5;
      aimag[i] := temp*exp(-hplus*ln(i-1));
   end;  (* for i *)

   (* generate y-values by applying fast Fourier
   transform -- the inverse FFT is the conjugate of the
   FFT multiplied by a scale factor, and we only need
   the real part of the inverse FFT *)

transform;
ymax := 0;  (* will hold maximum absolute value of y
   for scaling *)
for i := 1 to n do (* loop over x-values, in pixels *)
if abs(areal[i]) > ymax
   then ymax := abs(areal[i]);

   for i := 1 to n do
      y[i] := 90*areal[i]/ymax;

   (* plot the fractal *)
   graphcolormode;
   y[0] := y[n];
```

```
for i := 0 to n-1 do
    draw (i,100-round(y[i]),i+1,100-round(y[i+1]),3);
    (* line from ith to (i+1)th point *)

readln;
textmode(80);
writeln('Write data to file (Y/N) ?');
readln(flag);
if ((flag = 'Y') or (flag = 'y')) then
    begin
        writeln('Enter the name of the output file.');
        readln(filename);
        writeln('CAUTION : PROGRAM WILL
            OVERWRITE  ',filename);
        writeln('Continue (Y/N) ?');
        readln(flag);
        if ((flag = 'Y') or (flag= 'y')) then
            begin
                assign(outfile,filename);
                rewrite(outfile);
                for i := 1 to 256 do
                    writeln(outfile,y[i]);
                writeln(outfile,-9999);
                writeln(outfile,h);
                close(outfile);
            end;
    end;

end.
```

Mandelbrot–Weierstrass fractal surfaces (functions of two variables) are constructed similarly by generating appropriate two-dimensional spectra with 'amplitudes' given by the normal distribution

$$a_{m,n} = N(0, c/(H^{1+2m}H^{1+2n})) \qquad (12.4)$$

and random phases.

12.5.3 *Program for fractal surfaces (mw2dfractal)*

```
program mw2dfract;
    (* graphs Mandelbrot-Weierstrass fractals in two
    dimensions using (inverse) fast Fourier transform *)

uses
    crt,graph3;
```

```
var
    i, j, k, ihoriz, ivert : integer;
    h : real;                   (* Hurst exponent *)
    hplus : real;               (* Hurst exponent plus 1/2 *)
    terms : integer;            (* highest frequency in
                                spectrum, at most 15 *)
    areal, aimag : array[1..32,1..32] of real;
                                (* real and imag parts of
                                spectrum *)
    ymax : real;                (* maximum y to scale image *)
    scale : real;               (* to scale image *)
    temp : real;                (* temporary real *)
    twopidiv : real;            (* 2*pi / 32, so as to graph one
                                complete cycle in the 32 points
                                used *)

    index : array [1..32] of integer;  (* for bit reversal
                                in fft *)
    costable, sintable : array [0..32] of real;
                                (* for fft *)

    z : array [0..32,0..32] of real;  (* height on the
                                fractal surface *)
    zmax : real;

    u,v : array[0..32,0..32] of integer;  (* coordinates
                                for plotting on screen *)


const
    n = 32;                     (* size of fft *)
    hn = 16;                    (* half n *)
    log2n = 5;                  (* log to base 2 of n *)


procedure initbitrev; (* constructs index array for bit
    reversal in fft *)
    var k,l,lmax,ii,m : integer;
begin
    index[1] := 1;
    lmax := 1;
    ii := hn;

    for k := 1 to log2n do
        begin
            for l := 1 to lmax do
                index[l+lmax] := index[l] + ii;
                ii := ii div 2;
                lmax := 2 * lmax;
            end;
end;  (* procedure initbitrev *)
```

```
procedure trigtable;
    var
        k : integer;
        angle : real;

    begin
        for k := 0 to n do
            begin
                angle := 2*pi*k/n;
                costable[k] := cos(angle);
                sintable[k] := sin(angle);
            end;
    end;  (* procedure trigtable *)

procedure transform;
    var
        tempareal, tempaimag : array [1..32,1..32] of real;
        i0,i1,l,p, hp, ndivp : integer;
        c,s,r : real;
        t1, t2 : real;
    begin
        (* first do bit reversal, starting with vertical
            direction *)
        for ivert := 1 to n do
          begin
            i0 := index[ivert];
            for ihoriz := 1 to n do
                begin
                    tempareal[i0,ihoriz]
                        := areal[ivert,ihoriz];
                    tempaimag[i0,ihoriz]
                        := aimag[ivert,ihoriz];
                end;
          end;
        (* now continue with horizontal direction *)
        for ihoriz := 1 to n do
            begin
                i0 := index[ihoriz];
                for ivert := 1 to n do
                    begin
                        areal[ivert,i0]
                            := tempareal[ivert,ihoriz];
                        aimag[ivert,i0]
                            := tempaimag[ivert,ihoriz];
                    end;
            end;

        (* now do horizontal butterflies for fft *)
        p := 2;
        hp := 1;
        ndivp := n div p;
        for l := 1 to log2n do
```

```
begin
   for j := 1 to hp do
      begin
         c := costable[(j - 1) * ndivp];
         s := sintable[(j - 1) * ndivp];
         i := j;
         for i1 := 1 to ndivp do
            begin
               i0 := i + hp;
               for ivert := 1 to n do
                  begin
                     t1 := areal[ivert,i0] * c
                        - aimag[ivert,i0] * s;
                     t2 := areal[ivert,i0] * s
                        + aimag[ivert,i0] * c;
                     areal[ivert,i0]
                        := areal[ivert,i] - t1;
                     aimag[ivert,i0]
                        := aimag[ivert,i] - t2;
                     areal[ivert,i]
                        := areal[ivert,i] + t1;
                     aimag[ivert,i]
                        := aimag[ivert,i] + t2;
                  end;   (* end of ivert loop *)
               i := i + p;
            end;   (* end of i1 loop *)
      end;   (* end of j loop *)
   hp := p;
   p := p * 2;
   ndivp := ndivp div 2;
end;   (* end of l loop *)


(* now do vertical butterflies for fft *)
p := 2;
hp := 1;
ndivp := n div p;
for l := 1 to log2n do
   begin
      for j := 1 to hp do
         begin
            c := costable[(j - 1) * ndivp];
            s := sintable[(j - 1) * ndivp];
            i := j;
            for i1 := 1 to ndivp do
               begin
                  i0 := i + hp;

                  for ihoriz := 1 to n do
                     begin
                        t1 := areal[i0,ihoriz] * c
                           - aimag[i0,ihoriz] * s;
```

```
                 t2 := areal[i0,ihoriz] * s
                    + aimag[i0,ihoriz] * c;
                 areal[i0,ihoriz]
                    := areal[i,ihoriz] - t1;
                 aimag[i0,ihoriz]
                    := aimag[i,ihoriz] - t2;
                 areal[i,ihoriz]
                    := areal[i,ihoriz] + t1;
                 aimag[i,ihoriz]
                    := aimag[i,ihoriz] + t2;
              end;  (* end of ihoriz loop *)
           i := i + p;
        end;  (* end of i1 loop *)
      end;  (* end of j loop *)
hp := p;
p := p * 2;
         ndivp := ndivp div 2;
       end;  (* end of l loop *)

   end;  (* end transform *)

begin  (* main *)


   (* initialize *)
   twopidiv := 2*pi/32;  (* Turbo Pascal has pi as a
     predefined const. *)
   writeln('Enter the Hurst exponent H (0 to 1 (rough to
     smooth)).');
   readln(h);
   hplus := h + 0.5; (* to simplify calculating the
     spectrum *)
   writeln('Highest frequency (at most 15, 8 makes good
     mountains) ?');
   readln(terms);
   writeln('Input an integer (0 to 1000) to set random
     number generator.');
   readln(j);
   for i := 1 to j do
      temp := random;
   writeln('Program will draw fractal.  Type "enter" to
     clear');
   writeln('screen and exit when done.');


   (* setups for FFT *)
   initbitrev;
   trigtable;

   for i := 1 to n do
      for j := 1 to n do
         begin
             areal[i,j] := 0;
```

```
            aimag[i,j] := 0;
      end;   (* for j *)


(* generate the spectrum *)
for i := 2 to terms+1 do
   for j := 2 to terms + 1 do
      begin
         temp := 0;
         for k := 1 to 12 do
            temp := temp + random - 0.5;
         areal[i,j] := temp
            *exp(-hplus*ln((i-1)*(j-1)));
         temp := 0;
         for k := 1 to 12 do
            temp := temp + random - 0.5;
         aimag[i,j] := temp
            *exp(-hplus*ln((i-1)*(j-1)));
      end;   (* for j *)


(* generate z-values by applying fast Fourier transform
   -- the inverse FFT is the conjugate of the FFT
   multiplied by a scale factor, and we only need the
   real part of the inverse FFT *)

transform;
zmax := 0;   (* will hold maximum absolute value of z
   for scaling *)
for i := 1 to n do
   for j := 1 to n do (* loop over x-  and y-values, in
      pixels *)
      if abs(areal[i,j]) > zmax
         then zmax := abs(areal[i,j]);
for i := 1 to n do
   for j := 1 to n do
      z[i,j] := 20*areal[i,j]/zmax;
for j := 1 to 32 do   (* recovering the periodicity *)
   z[0,j] := z[32,j]; (* of the trig. functions or *)
for i := 0 to 32 do    (* complex exponentials *)
   z[i,0] := z[i,32]; (* used to build the fractal *)


(* plot the fractal *)
graphcolormode;
for i := 0 to 32 do
   for j := 0 to 32 do
      begin
         (* calculate coordinates for plotting on
            screen *)
         (* Vectors map from 3d to 2d as follows:
            3d coordinates are (i,j,z[i,j])
            (1,0,0) --> (-3,4)  (vertical coord.
```

```
             increases downward)
        (0,1,0) --> (5,0)

        (0,0,1) --> (0,1).
        A translation is then added. *)

      u[i,j] := round(-3*i + 5*j    + 130);
      v[i,j] := round( 4*i - z[i,j] + 20);
   end;


for i := 0 to 31 do
   for j := 0 to 32 do
      draw(u[i,j],v[i,j],u[i+1,j],v[i+1,j],3);
for i := 0 to 32 do
   for j := 0 to 31 do
      draw(u[i,j],v[i,j],u[i,j+1],v[i,j+1],3);

readln;
textmode(80);

end.
```

Finally, Mandelbrot–Weierstrass fractals can be considered as 'waves upon waves'. Figure 12.2 illustrates this construction by varying the number of terms used to construct the Mandelbrot–Weierstrass fractals in the programs above.

12.6 Fractal analysis of patterns

We develop illustrative programs for the computation of fractal exponents. We begin with a module common to all of the programs, which we call the *generic hyperbolic distribution fitter*. This program fits a hyperbolic distribution to a set of data represented as two arrays of points $u[n]$ and $v[n]$. The data is first log-transformed in place to reduce the problem to that of fitting a straight line. The intercept a and slope b of that line are computed using standard linear regression techniques. The coefficient of correlation ρ (r in the program) is also computed.

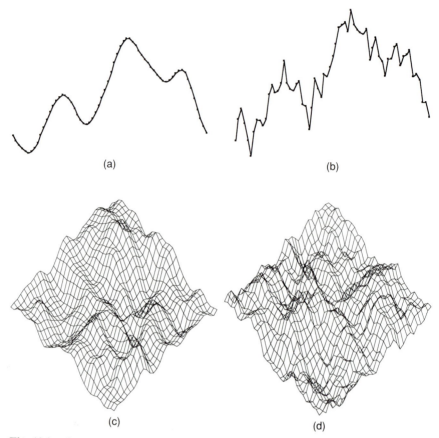

Fig. 12.2 Constructing a Mandelbrot–Weierstrass fractal. Top one-dimensional Mandelbrot–Weierstrass fractal with Hurst exponent 0.5. (a) Approximation with 8 frequency components. (b) Approximation with 31 frequency components. Bottom: two-dimensional Mandelbrot–Weierstrass fractal with Hurst exponent 0.5. (c) Approximation with 4 frequency components in each direction (16 in all). (d) Approximation with 15 frequency components in each direction (225 in all).

12.6.1 *Generic hyperbolic distribution fitter program (demo)*

```
program demo;
     (* demonstrates linear regression of log-transformed
     data, up to 1000 data points *)

uses
     printer;
```

```
type
    realarray = array [1..1024] of real;
    fractname = string[24];    (* name of exponent *)


var
    xx, yy : realarray;    (* input arrays *)
    xvalue : real;
    k : integer;           (* number of points *)
    fractfactor : real;    (* convert slope to fractal
                              exponent *)
    filename : string[16];
    infile : text;
    printflag : char;      (* whether to send output to
                              printer *)



procedure logregress(x,y : realarray; k : integer;
    expfactor : real; expname : fractname);

    (* Computes the fractal exponent expname given two
    arrays x[i], y[i] of real numbers, indexed on
    1,2,...,k, and a factor expfactor.  First, linear
    regression is used to fit log-transformed data to the
    equation  ln(y) = a + b ln(x), yielding the slope  b
    and the coefficient of correlation  r.  The fractal
    exponent is expfactor times the slope. *)

var
    sumx : real;           (* sum of x-values *)
    sumy : real            (* sum of y-values *)
    sumxsq : real;         (* sum of x**2      *)
    sumysq : real;         (* sum of xy        *)
    sumxy : real;          (* sum of y**2      *)
    a : real;              (* intercept of regression
                             line *)
    b : real;              (* slope of regression line *)
    sdb : real;            (* standard deviation of b *)
    r : real;              (* coeffcient of correlation *)
    xbar,ybar : real;      (* average x, average y *)
    i : integer;

begin
    sumx := 0;
    sumy := 0;
    sumxy := 0;
    sumxsq := 0;
    sumysq := 0;

    for i := 1 to k do
      begin
        sumx :=  sumx + ln(x[i]);
```

```
      sumy :=  sumy + ln(y[i]);
      sumxy := sumxy + ln(x[i])*ln(y[i]);
      sumxsq := sumxsq + ln(x[i])*ln(x[i]);
      sumysq := sumysq + ln(y[i])*ln(y[i]);
   end;


xbar := sumx/k ;
ybar := sumy/k ;

b := (sumxy - k*xbar*ybar)/(sumxsq - k*xbar*xbar);

a := ybar - (b*xbar);

r := (sumxy - k*xbar*ybar)/sqrt((sumxsq - k*xbar*xbar)
          *(sumysq - k*ybar*ybar));



writeln(expname,'   ',expfactor*b);
if printflag = 'Y' then
   writeln(lst,expname,'   ',expfactor*b);

if k > 4 then
   begin
      if ((r*r) < 1) then    (* to avoid problems if r
                              computes to 1+epsilon *)
         sdb := sqrt((1-r*r)*(sumysq-k*ybar*ybar)/
            ((k-4)*(sumxsq-k*xbar*xbar)))
      else
         sdb := 0;
      writeln('+/-  ',expfactor*sdb);
      if k < 27 then
         writeln('Use t-test with ',k-2,
            ' degrees of freedom.');
      end
      else (* k <= 4 *)
         begin
            write;
            writeln('standard deviation undefined');
         end;
   writeln('coeff of correl.  ',r);

   if printflag = 'Y' then
      begin   (* output to printer *)
         if k > 4 then
            begin
              writeln(lst,'+/-  ',expfactor*sdb);
              if k < 27 then
                 writeln(lst,'Use t-test with ',k-2,
                    ' degrees of freedom.');
              end
```

```
          else (* k <= 4 *)
              begin
                  write(lst);
                  writeln(lst,'standard deviation
                      undefined');
              end;
          writeln(lst,'coeff of correl.  ',r);
      end;

end;  (* logregress *)

begin (* main *)

    writeln('Do you want output to printer (Y/N)');
    readln(printflag);
    if printflag = 'y' then
        printflag := 'Y';
    if printflag ='Y' then
        writeln('Please turn on the printer.');
    writeln('Data must be in the form   x  y   on each
            line.');
    writeln('-1 - 1 denotes end of data.');
    writeln('Type the name of the input data file.');
    writeln('Type "none" to enter data from keyboard');


    (* read in the data *)

    xvalue := -1;          (* -1 is flag for end of file *)
readln(filename);
if filename = 'none' (* data entered from keyboard *)
then
    begin
        writeln('Enter data in the form   x  y   on
            each line.');
        writeln('Enter -1 -1 at end of data.');

        k := 1;
        repeat
            readln(xx[k],yy[k]);
            xvalue := xx[k];
            if xvalue > 0 then k := k+1;  (* if xx[k]
                    > 0, then increment k *)
            if k > 1002 then
              begin
                  writeln('TOO MUCH DATA.');
                  halt;
              end;
        until(xvalue < 0);       (* negative value
                                 marks end of data *)
          k := k-1;              (* making k the number
                                 of points *)
```

```
        end

else
    begin                    (* data from input file *)
        if printflag = 'Y' then
            write(lst,filename,'   ');
        assign(infile,filename);
        reset(infile);


        k := 1;
        repeat
            readln(infile,xx[k],yy[k]);
            xvalue := xx[k];
            if xvalue > 0 then k := k+1;   (* if xx[k]
                > 0, then increment k *)
            if k > 1002 then
              begin
                writeln('FILE TOO LONG.');
                close(infile);
                halt;
              end;
        until(xvalue < 0);    (* negative value

                                marks end of data *)
                close(infile);
                k := k-1;            (* making k the number
                                     of points *)
        end;


    logregress(xx,yy,k,1,'exponent   ');

end.   (* main *)
```

The computation of the area–perimeter exponent and consequent fractal dimension, the box dimension and the cluster dimension, is similar. Programs for the area–perimeter exponent and the cluster dimension follow.

12.6.2 *Area–perimeter exponent program (areaperi)*

```
program areaperi;
        (* Computes the fractal dimension of the boundaries
    of a family of self-similar subsets in the plane.
    The dimension is computed as twice the area-perimeter
    exponent. The input data consists up to 1000 ordered
    pairs(area[i], perimeter[i]), one pair for each
    object. *)
```

```
uses
    printer;

type
    realarray = array [1..1024] of real;
    fractname = string[24];

var
    xx, yy : realarray;       (* area, perimeter *)
    xvalue : real;
    k : integer;              (* number of points *)
    fractfactor : real;       (* converts slope to fractal
                                 exponent *)
    filename : string[80];
    infile : text;
    printflag : char;         (* whether to send output to
                                 printer *)

procedure logregress(x,y : realarray; k : integer;
        expfactor : real; expname : fractname);

        (* Computes the fractal exponent expname given two
        arrays x[i], y[i] of real numbers, indexed on
        1,2,...,k, and a factor expfactor.  First, linear
        regression is used to fit log-transformed data to the
        equation  ln(y) = a + b ln(x), yielding the slope  b
        and the coefficient of correlation  r.  The fractal
        exponent is expfactor times the slope. *)

var
    sumx : real;              (* sum of x-values *)
    sumy : real;              (* sum of y-values *)
    sumxsq : real;            (* sum of x**2     *)
    sumysq : real;            (* sum of xy       *)
    sumxy : real;             (* sum of y**2     *)
    a : real;                 (* intercept of regression
                                 line *)
    b : real;                 (* slope of regression line *)
    sdb : real;               (* standard deviation of b *)
    r : real;                 (* coefficient of correlation *)
    xbar,ybar : real;         (* average x, average y *)
    i : integer;

begin
    sumx := 0;
    sumy := 0;
    sumxy := 0;
    sumxsq := 0;
    sumysq := 0;

    for i := 1 to k do
      begin
```

```
        sumx :=  sumx + ln(x[i]);
        sumy :=  sumy + ln(y[i]);
        sumxy := sumxy + ln(x[i])*ln(y[i]);
        sumxsq := sumxsq + ln(x[i])*ln(x[i]);
        sumysq := sumysq + ln(y[i])*ln(y[i]);
      end;


xbar := sumx/k ;
ybar := sumy/k ;


b := (sumxy - k*xbar*ybar)/(sumxsq - k*xbar*xbar);

a := ybar - (b*xbar);

r := (sumxy - k*xbar*ybar)/sqrt((sumxsq - k*xbar*xbar)
          *(sumysq - k*ybar*ybar));

writeln(expname,'  ',expfactor*b);
if printflag = 'Y' then
    writeln(lst,expname,'   ',expfactor*b);


if k > 4 then
    begin
        if ((r*r) < 1) then    (* to avoid problems if r
                              computes to 1+epsilon *)
            sdb := sqrt((1-r*r)*(sumysq-k*ybar*ybar)/
               ((k-4)*(sumxsq-k*xbar*xbar)))
        else
            sdb := 0;
        writeln('+/-  ',expfactor*sdb);
        if k < 27 then
            writeln('Use t-test with ',k-2, ' degrees of
               freedom.');
    end
    else (* k <= 4 *)
        begin
            write;
            writeln('standard deviation undefined');
        end;
writeln('coeff of correl.  ',r);


if printflag = 'Y' then
    begin   (* output to printer *)
        if k > 4 then
          begin
            writeln(lst,'+/-  ',expfactor*sdb);
            if k < 27 then
              writeln(lst,'Use t-test with ',k-2,
                ' degrees of freedom.');
          end
          else (* k <= 4 *)
```

```
              begin
                    write(lst);
                    writeln(lst,'standard deviation
                      undefined');
              end;
          writeln(lst,'coeff of correl.  ',r);
      end;


end;  (* logregress *)


(*******************)


begin (* main *)


      writeln('Do you want output to printer (Y/N)');
      readln(printflag);
      if printflag = 'y' then
         printflag := 'Y';
      if printflag ='Y' then
         writeln('Please turn on the printer.');
      writeln('Type the name of the input data file.');
      writeln('Type "none" to enter data from keyboard');

      writeln('File must be in the form   area perimeter
            on each line.');

      (* read in the data *)
      xvalue := 1;                    (* -9999 is flag for end
                                          of file *)
      readln(filename);

      if filename = 'none'        (* data entered from
                                       keyboard *)
      then
          begin
              writeln('Enter the data, one point at a time,
                  in the form');
              writeln('area  perimeter,  followed by
                  "enter"');
              writeln('Enter at most 1000 points.');
              writeln('Enter -9999 -9999 as flag for end of
                  data.');
              k := 1;
              repeat
                 readln(xx[k],yy[k]);
                 xvalue := xx[k];
                 if xvalue > 0 then k := k+1;  (* if xx[k]
                       > 0,then increment k *)
                 if k > 1002 then
```

```
                begin
                    writeln('TOO MUCH DATA.');
                    halt;
                end;
            until(xvalue < 0);        (* negative value
                                         marks end of data *)
                k := k-1;             (* making k the number
                                         of points *)
        end

    else
        begin                              (* data from input
                                              file *)
            if printflag = 'Y' then
                write(lst,filename,'   ');
            assign(infile,filename);
            reset(infile);

            k := 1;
            repeat
                    readln(infile,xx[k],yy[k]);
                    xvalue := xx[k];
                    if xvalue > 0 then k := k+1;   (* if xx[k]
                        > 0,then increment k *)
                    if k > 1002 then
                      begin
                        writeln('FILE TOO LONG.');
                        close(infile);
                        halt;
                      end;
                until(xvalue < 0);    (* negative value
                                         marks end of data *)
                    close(infile);
                    k := k-1;             (* making k the number
                                             of points *)
            end;
        logregress(xx,yy,k,2,'dim of boundary.');

end.   (* main *)
```

12.6.3 *Cluster dimension program (clusterd)*

```
program clusterd;
    (* computes cluster (correlation) dimension of a set
    of up to 1000 points in the plane *)

uses
    printer;
```

```
type
    realarray = array[1..1024] of real;
    fractname = string[24];


var
    xx,yy : realarray;        (* input data *)
    xvalue : real;            (* flag for end of file *)
    i,j,k,l,i1,j1,tempint : integer;
    infile : text;
    filename : string[80];
    number, lognumber : array[-32..31] of real; (* number
                               of points within a distance
                               2**(i/2) of a typical point,
                               and log of that number *)

    temp : real;
    small,big: integer;       (* bounds for scaling region *)
    rptflag : string[2];      (* to repeat regression over
                               a different scaling region *)

    printflag : char;         (* whether to send output to
                               printer *)


procedure regress(small,big : integer; expfactor : real;
     expname : fractname);


      (* Modified from logregress procedure in other
      programs for fractal exponents. *)

      (* Computes the fractal exponent "cluster dimension"
      from an array distance[n] containing the cumulative
      number of points within a distance 2**(n/2) of a
      typical point.
           First, linear regression is used to fit the data
      to the equation  lognumber = a + bn, yielding the
      slope  b  and the coefficient of correlation  r.
      The fractal exponent is expfactor (-1) times the
      slope.    *)

var

        sumx : real;              (* sum of x-values *)
        sumy : real;              (* sum of y-values *)
        sumxsq : real;            (* sum of x**2      *)
        sumysq : real;            (* sum of xy        *)
        sumxy : real;             (* sum of y**2      *)
        a : real;                 (* intercept of regression
                                  line *)
        b : real;                 (* slope of regression line *)
        sdb : real;               (* standard deviation of b *)
        r : real;                 (* coefficient of correlation *)
        xbar,ybar : real;         (* average x, average y *)
        i : integer;
```

```
begin
    sumx := 0;
    sumy := 0;
    sumxy := 0;
    sumxsq := 0;
    sumysq := 0;
    for i := small to big do
        begin
            sumx :=  sumx + i;
            sumy :=  sumy + lognumber[i];
            sumxy := sumxy + i*lognumber[i];
            sumxsq := sumxsq + i*i;
            sumysq := sumysq + lognumber[i]*lognumber[i];
        end;


    k := big-small + 1;
    xbar := sumx/k ;
    ybar := sumy/k ;


    b := (sumxy - k*xbar*ybar)/(sumxsq - k*xbar*xbar);

    a := ybar - (b*xbar);

    r := (sumxy - k*xbar*ybar)/sqrt((sumxsq - k*xbar*xbar)
                *(sumysq - k*ybar*ybar));



    writeln(expname,'   ',expfactor*b);
    if printflag = 'Y' then
        writeln(lst,expname,'   ',expfactor*b);

    if k > 4 then
begin
    if ((r*r) < 1) then (* to avoid problems if r
            computes to 1+epsilon *)
        sdb := sqrt((1-r*r)*(sumysq-k*ybar*ybar)/
            ((k-4)*(sumxsq-k*xbar*xbar)))
    else
        sdb := 0;
    writeln('+/-  ',expfactor*sdb);
    if k < 27 then
        writeln('Use t-test with ',k-2, ' degrees of
            freedom.');
end
else (* k <= 4 *)
    begin
        write;
        writeln('standard deviation undefined');
    end;
writeln('coeff of correl.  ',r);
```

```
   if printflag = 'Y' then
      begin     (* output to printer *)
         if k > 4 then
           begin
             writeln(lst,'+/-  ',expfactor*sdb);
             if k < 27 then
               writeln(lst,'Use t-test with ',k-2,
                 ' degrees of freedom.');
           end
           else (* k <= 4 *)
              begin
                   write(lst);
                   writeln(lst,'standard deviation
                     undefined');
              end;
         writeln(lst,'coeff of correl.  ',r);
      end;


end;   (* regress *)

(**************)

begin (* main *)

      writeln('Do you want output to printer (Y/N)');
      readln(printflag);
      if printflag = 'y' then
         printflag := 'Y';
      if printflag ='Y' then
         writeln('Please turn on the printer.');
      writeln('Type the name of the input data file.');
      writeln('Type "none" to enter data from keyboard.');
      readln(filename);

      (* read in the data *)

      xvalue := 1;          (* a value of -9999 will be a
                               flag to stop *)


      if filename = 'none' (* data entered from keyboard *)
      then
           begin
               writeln('Enter the data, one point at a time,
                   in the');
               writeln('format  100 200.  Follow each point
                   with "enter".');
               writeln('Enter -9999 -9999 to indicate end of
                   data.  Enter');
               writeln('at most 1000 points');
```

```
            k := 1;
            repeat
               readln(xx[k],yy[k]);
               xvalue := xx[k];
               if xvalue > -9998 then k := k+1;
               if k > 1002 then
                 begin
                    writeln('TOO MUCH DATA.');
                    halt;
                 end;
              until(xvalue < -9998);  (* marks end of
                                         data *)
              k := k-1;              (* making k the number
                                         of points *)
        end

  else
        begin                       (* data from input file *)

             if printflag = 'Y' then
                write(lst,filename,'   ');

             assign(infile,filename);
             reset(infile);

             k := 1;
             repeat
                readln(infile,xx[k],yy[k]);
                xvalue := xx[k];
                if xvalue > -9998 then k := k+1;
                if k > 1002 then
                  begin
                     writeln('FILE TOO LONG.');
                     close(infile);
                     halt;
                  end;
               until(xvalue < -9998);   (* marks end of
                                          data *)
               close(infile);
               k := k-1;               (* making k the
                                          number of points *)
        end;
        writeln(k,'  points');

  for i := -32 to 31 do
     begin
          number[i] := 0;
          lognumber[i] := -999999.0;
     end;
  for i := 1 to k do
    for j := 1 to i-1 do
```

```
begin
  temp := (xx[i]-xx[j])*(xx[i]-xx[j]) + (yy[i]
          -yy[j])*(yy[i]-yy[j]);
  if temp > 0 then
    begin
      temp := ln(temp)/(ln(2));      (* thus log base
            2 of distance squared or log to base
            radical 2 of distance *)
      l := trunc(temp);
      if ((l >= -32) and (l <= 31)) then
          number[l] := number[l]+2/k;
    end;
end;


writeln('Here is the number of points within each
 distance class');
writeln('of a typical point.  Distance class  i   means
 within');
writeln('a distance  2**(i/2).  Note which classes are
 occupied');
writeln('to determine scaling region.');
writeln('The data will appear on several screens and
 on the printer.');
writeln('Type "enter" to continue after you have read
 each screen.');
writeln('size class, number');
if printflag = 'Y' then
  writeln(lst,'size class, number');


for i := -31 to -10 do
  begin
        number[i] := number[i] + number[i-1];
        if ((number[i] > 0) and ((number[i]
            <>number[i-1])
          or (number[i-1] < k-1.9))) then
             (* if size class is not empty, and system
             has not saturated *)
          begin
             writeln(i,' ',number[i]);
             if printflag = 'Y' then
                writeln(lst,i,' ',number[i]);
             lognumber[i] := 2*ln(number[i])/(ln(2));
                     (* thus log to base radical 2 *)
          end;
  end;
writeln('Type "enter" to continue.');
readln;  (* to continue *)
for i := -9 to 11 do

  begin
        number[i] := number[i] + number[i-1];
```

```
      if ((number[i] > 0) and ((number[i]
         <>number[i-1])
       or (number[i-1] < k-1.9))) then
      begin
          writeln(i,' ',number[i]);
          if printflag = 'Y' then
             writeln(lst,i,'   ',number[i]);
          lognumber[i] := 2*ln(number[i])/(ln(2));
                  (* thus log to base radical 2 *)
      end;
end;
   writeln('Type "enter" to continue.');
   readln;  (* to continue *)
   for i := 12 to 32 do
      begin
          number[i] := number[i] + number[i-1];
          if ((number[i] > 0) and ((number[i]
             <>number[i-1])
           or (number[i-1] < k-1.9))) then
          begin
              writeln(i,' ',number[i]);
              if printflag = 'Y' then
                 writeln(lst,i,'   ',number[i]);
              lognumber[i] := 2*ln(number[i])/(ln(2));
                    (* thus log to base radical 2 *)
          end;
      end;

   rptflag := 'Y';
   repeat
      writeln('Enter the bounds for the regession, for
         example, 4   11.');
      writeln('The the lower bound must be less than the
         upper bound.');
      readln(small,big);
      if printflag = 'y' then
         writeln(lst,'limits of regression:', small,
         ' to ', big);
      regress(small,big,1,'cluster dim.');

      writeln('Do you want to change the limits and
         repeat (Y/N) ?');
      readln(rptflag);
   until((rptflag = 'N') or (rptflag ='n'));
end.
```

One can easily construct a test version of the program *clusterdim* by replacing the file input above by a module to generate points randomly in the plane such as the following program fragment:

```
readln(j);      (*number of points*)
for i := 1 to j do
     begin
          x[j] := random;
          y[j] := random;
     end;
```

Such test versions serve several purposes. Since a set of points randomly chosen from a uniform distribution on the plane has expected 'experimental' cluster dimension 2, the test version can test the computer code for the cluster dimension and the random number generator used to generate the data (cf. Marsaglia 1968; Knuth 1981). Moreover, programs like this can generate confidence limits for null hypotheses, like the hypothesis that a set of points is a uniform random subset of the plane. See Section 8.3 and Hastings *et al.* (1992) for applications.

12.7 Fractal exponents for time series

The following programs illustrate the techniques described in Chapter 4 for computing the Hurst exponent of a time series. We begin with the growth of range and growth of second moment techniques.

12.7.1 *Program for growth of range and second moment (hurst)*

```
program Hurst;
        (*  Computes the Hurst exponent of time series
        using the growth of range and growth of second moment
        algorithms. *)

uses
    printer;

type
    realarray = array[1..1024] of real;
    fractname = string[24];

var
    filename : string[80];
    infile : text;
    i,j,k,l,m : integer;
    x : realarray;          (* for input data *)
    xvalue: real;           (* to test for end of file *)
```

```
    mom2, range : realarray;   (* second moment, range, over
                                  specified lag *)
    minx,maxx : real;          (* minimum and maximum values
                                  over specified lag *)
    printflag : char;          (* whether to send output to
                                  printer *)


procedure logregress(y : realarray; k : integer;
    expfactor : real; expname : fractname);

    (* Computes the fractal exponent expname the array
    y[i] of real numbers, indexed on 1,2,...,k, and a
    factor expfactor.  First, linear regression is used
    to fit log-transformed data to the equation  ln(y)
    = a + b ln(i), yielding the slope  b  and the
    coefficient of correlation  r.  The fractal exponent
    is expfactor times the slope. *)

var
    sumx : real;               (* sum of x-values *)
    sumy : real;               (* sum of y-values *)
    sumxsq : real;             (* sum of x**2     *)
    sumysq : real;             (* sum of xy       *)
    sumxy : real;              (* sum of y**2     *)
    a : real;                  (* intercept of regression
                                  line *)
    b : real;                  (* slope of regression line *)
    sdb : real;                (* standard deviation of b *)
    r : real;                  (* coefficient of correlation *)
    xbar,ybar : real;          (* average x, average y *)
    i : integer;

begin
    sumx := 0;
    sumy := 0;
    sumxy := 0;
    sumxsq := 0;
    sumysq := 0;

    for i := 1 to k do
      begin
        sumx :=  sumx + ln(i);
        sumy :=  sumy + ln(y[i]);
        sumxy := sumxy + ln(i)*ln(y[i]);
        sumxsq := sumxsq + ln(i)*ln(i);
        sumysq := sumysq + ln(y[i])*ln(y[i]);
      end;

    xbar := sumx/k ;
    ybar := sumy/k ;
```

```
    b := (sumxy - k*xbar*ybar)/(sumxsq - k*xbar*xbar);

    a := ybar - (b*xbar);

    r := (sumxy - k*xbar*ybar)/sqrt((sumxsq - k*xbar*xbar)
            *(sumysq - k*ybar*ybar));

    writeln(expname,'   ',expfactor*b);
    if printflag = 'Y' then
        writeln(lst,expname,'   ',expfactor*b);

    if k > 4 then
        begin
            if ((r*r) < 1) then (* to avoid problems if r
                    computes to 1+epsilon *)
                sdb := sqrt((1-r*r)*(sumysq-k*ybar*ybar)/
                    ((k-4)*(sumxsq-k*xbar*xbar)))
            else
                sdb := 0;
            writeln('+/-  ',expfactor*sdb);
            if k < 27 then
                writeln('Use t-test with ',k-2, ' degrees of
                    freedom.');
        end
        else (* k <= 4 *)
            begin
                write;
                writeln('standard deviation undefined');
            end;
    writeln('coeff of correl.  ',r);

    if printflag = 'Y' then
        begin   (* output to printer *)
            if k > 4 then
             begin
                writeln(lst,'+/-  ',expfactor*sdb);
                if k < 27 then
                    writeln(lst,'Use t-test with ',k-2,
                      ' degrees of freedom.');
                end
                else (* k <= 4 *)
                    begin
                        write(lst);
                        writeln(lst,'standard deviation
                            undefined');
                    end;
            writeln(lst,'coeff of correl.  ',r);
        end;

end;   (* logregress *)

(***************)
```

```
    begin  (* main *)

        writeln('Do you want output to printer (Y/N)');
        readln(printflag);
        if printflag = 'y' then
            printflag := 'Y';
        if printflag ='Y' then
            writeln('Please turn on the printer.');
        writeln('Type the name of the input data file.');
        writeln('Type "none" to enter data from keyboard.');

(* read in data *)
xvalue := 1;    (* -9999 is flag for end of file *)
readln(filename);
if filename = 'none' (* data entered from keyboard *)
then
    begin
        writeln('Enter the time series, one point at
          a time.');
        writeln('Follow each point with the "enter"
          key.');
        writeln('Number of points must be at most
          1000.');
        writeln('Enter  -9999  to denote the end of
          data.');
        j := 1;
        repeat
            readln(xvalue);
            if xvalue > -9999 then
                begin
                    x[j] := xvalue;
                    j := j+1;
                end;
            if j > 1002 then
                begin
                    writeln('Too many points');
                    halt;
                end;
        until xvalue < -9998;
        j := j-1;

        writeln('Length of time series = ',j);
    end

else
    begin (* data from input file *)
        if printflag = 'Y' then
            writeln(lst,filename,'  ');
        assign(infile,filename);
        reset(infile);

        j := 1;
```

```
        repeat
               readln(infile,xvalue);
               if xvalue > -9999 then
                   begin
                       x[j] := xvalue;
                       j := j+1;
                   end;
                 if j > 1002 then
                   begin
                           writeln('File too long');
                           close(infile);
                           halt;
                   end;
          until xvalue < -9998;
          j := j-1;
          close(infile);
          writeln('Length of file = ',j);

   end; (* module for file input *)
   writeln('enter the maximum lag (preferably at
    most 1/4 the');
   writeln('number of data points)');
   readln(k);

(* calculate *)
for i := 1 to k do                    (* vary lag    i *)
    begin
        mom2[i] := 0;
        range[i] := 0;

        for l := 1 to j-i do   (* vary starting point
               l *)
            begin
            (* for growth of moment *)
            mom2[i] := mom2[i] + (x[l+i]
               -x[l])*(x[l+i]-x[l]);

            (* for growth of range *)
            minx := x[l];
            maxx := x[l];
            for m := l+1 to l+i do
                 begin
                        if x[m] < minx then
                            minx := x[m];
                        if x[m] > maxx then
                            maxx := x[m];
                 end;
             range[i] := range[i] + maxx - minx;
             end;

         mom2[i] := mom2[i]/(j-i);
         range[i] := range[i]/(j-i);
    end;
```

```
        logregress(mom2,k,0.5,'H from 2nd moment ');
        logregress(range,k,1,'H from range ');
        writeln('See text, Section 4.4, for cautions on use of
               growth');
        writeln('of range ');
        if printflag= 'Y' then
            writeln(lst,'See text, Section 4.4, for cautions on
               use of growth');
        if printflag ='Y' then
            writeln(lst,'of range ');


end.
```

One can compute a local Hurst exponent by computing the expectation of the product of two successive spatial increments as well as their respective second moments, and then compute the fractal dimension.

12.7.2 *Program for local growth of second moment (localH)*

```
program localH;
        (*  Computes the Hurst exponent of time series
        using the local growth of second moment algorithm. *)

uses
    printer;

var
    filename : string[80];
    infile : text;
    i,j,k : integer;
    x : array [1..1024] of real;
    xvalue: real;
    lag : integer;
    deltax1, deltax2 : real;    (* increments of time
                        series *)
    mom1 : real;        (* 2nd moment of first increment *)
    mom2 : real;        (* 2nd moment of second increment *)
    expprod : real;     (* expectation of product of
                        increments *)
    rho : real;
    H : real;
    printflag : char;       (* whether to send output to
                        printer *)

begin

        writeln('Do you want output to printer (Y/N)');
        readln(printflag);
```

```
if printflag = 'y' then
   printflag := 'Y';
if printflag ='Y' then
   writeln('Please turn on the printer.');
writeln('Type the name of the input data file.');
writeln('Type "none" to enter data from keyboard');
readln(filename);

(* read in data *)
xvalue := 1; (* -9999 is flag for end of file *)

if filename = 'none' (* data entered from keyboard *)
  then
    begin
        writeln('Enter the time series, one point at
          a time.');
        writeln('Follow each point with the "enter"
          key.');
        writeln('Number of points must be at most
          1000.');
        writeln('Enter  -9999  to denote the end of
          data.');
        j := 1;
        xvalue := 1;
        repeat
              readln(xvalue);
              if xvalue > -9999 then
                  begin
                     x[j] := xvalue;
                     j := j+1;
                  end;
              if j > 1002 then
                  begin
                         writeln('Too many points');
                         halt;
                  end;
        until xvalue < -9998;
        j := j-1;

        writeln('Length of time series = ',j);
    end

else
    begin  (* data from input file *)
        if printflag = 'Y' then
            writeln(lst,filename,'   ');
        assign(infile,filename);
        reset(infile);

        j := 1;
        xvalue := 1;
```

```
repeat
        readln(infile,xvalue);
        if xvalue > -9999 then
            begin
                x[j] := xvalue;
                j := j+1;
            end;
        if j > 1002 then
            begin
                    writeln('File too long');
                        close(infile);
                        halt;
                end;
        until xvalue < -9998;
        j := j-1;
        close(infile);
        writeln('Length of file = ',j);

end; (* module for file input *)

writeln('Input the maximum time interval.');
writeln('At most 1/4 the length of the file.');
readln(k);

lag := 1;

writeln('lag    relative      H');
writeln('        product of');
writeln('        successive');
writeln('        increments');

if printflag = 'Y' then
  begin
        writeln(lst,'lag    relative      H');
        writeln(lst,'        product of');
        writeln(lst,'        successive');
        writeln(lst,'        increments');
  end;

repeat

  (* compute product of successive increments *)
  (* intialize *)

    mom1 := 0;        (* 2nd moment of first
                        increment *)
    mom2 := 0;        (* 2nd moment of second
                        increment *)
    expprod := 0;     (* expectation of product
                        of increments *)

    for i := 1 to j - (2*lag) do  (* vary
                        starting point *)
```

```
begin
  deltax1 := x[i + lag] - x[i];
                    (* first incr. *)
  deltax2 := x[i + lag + lag]
    - x[i + lag];   (* second incr. *)
  mom1 := mom1 + deltax1*deltax1;
  mom2 := mom2 + deltax2*deltax2;
  expprod := expprod + deltax1*deltax2;
end;

rho := expprod/sqrt(mom1*mom2);
                    (* caution: the coefficient
                    of correlation only in the
                    case of zero expectations
                    of increments *)
if rho > -1 then
  begin
    H := ln(2+2*rho)/ln(4);
                    (* 2**2H = 2+2rho *)
    writeln(lag,'       ',rho:8:6,
        '   ',H:9:6);
    if printflag = 'Y' then
      writeln(lst,lag,'       ',rho:8:6,
        '   ',H:9:6);
  end
else
    begin
      writeln('cannot compute H');
      if printflag = 'Y' then
        writeln(lst,'cannot compute H');
      halt;
    end;
  lag := 2 * lag;

until lag > k;

end.
```

The final program in this section, *pwrspect*, obtains the Hurst exponent of a time series using Fourier transform methods from Section 4.6 and Chapter 5.

12.7.3 *Program for Hurst exponent from power spectrum (pwrspect)*

```
program pwrspect;
    (* Computes Hurst exponent of a time series of
    256 points using FFT methods *)
```

```
uses
    printer;


type
    realarray = array[1..1024] of real;

var
    filename : string[80];
    infile : text;
    i,j,k,kk,l,m : integer;
    xvalue: real;
    areal, aimag : array[1..256] of real;
    index : array [1..256] of integer;   (* for bit reversal
                        in fft *)
    costable, sintable : array [0..256] of real;   (* for
                        fft *)
    energy : realarray;       (* entry in power spectrum *)
    median : array[0..2] of real;    (* median filter on
                        energy before regression *)
    filtflag : char;          (* whether to apply filter *)
    temp : real;              (* temporary *)
    printflag : char;         (* whether to send output to
                        printer *)


const
    n = 256;          (* number of terms, size of fft *)
    hn = 128;         (* half n *)
    log2n = 8;        (* log to base 2 of n *)


procedure initbitrev; (* constructs index array for bit
    reversal in fft *)
    var k,l,lmax,ii,m : integer;

    begin
        index[1] := 1;
        lmax := 1;
        ii := hn;

        for k := 1 to log2n do
            begin
                for l := 1 to lmax do
                    index[l+lmax] := index[l] + ii;
                    ii := ii div 2;
                    lmax := 2 * lmax;
                end;
        end;   (* procedure initbitrev *)

procedure trigtable;
    var
        k : integer;
        angle : real;
```

```
    begin
        for k := 0 to n do
            begin
                angle := 2*pi*k/n;
                costable[k] := cos(angle);
                sintable[k] := sin(angle);
            end;
    end;  (* procedure trigtable *)

procedure transform;
    var
        i,i0,i1,j,l,p,hp, ndivp : integer;
        tempreal, tempimag : array[1..256] of real;
        c,s : real;   (* cosine, sine *)
        t1, t2 : real;  (* temporaries *)

    begin
        (* first do bit reversal *)
        for i := 1 to n do
            begin
                i0 := index[i];
                tempreal[i] := areal[i0];
                tempimag[i] := aimag[i0];
            end;
        for i := 1 to n do
            begin
                areal[i] := tempreal[i];
                aimag[i] := tempimag[i];
            end;

        (* initialize some variables *)
        p := 2;
        hp := 1;
        ndivp := n div p;

        for l := 1 to log2n do
            begin
                for j := 1 to hp do
                    begin
                        c := costable[(j-1)*ndivp];
                        s := sintable[(j-1)*ndivp];
                        i := j;
                        for i1 := 1 to ndivp do
                            begin
                                i0 := i + hp;
                                t1 := areal[i0]*c - aimag[i0]*s;
                                t2 := areal[i0]*s + aimag[i0]*c;
                                areal[i0] := areal[i] - t1;
                                aimag[i0] := aimag[i] - t2;
                                areal[i] := areal[i] + t1;
                                aimag[i] := aimag[i] + t2;
                                i := i + p;
```

```
                        end;   (* for il *)
                  end;   (* for j *)
            hp := p;
            p := 2*p;
            ndivp := ndivp div 2;
            end;    (* for l *)

end;   (* procedure transform *)


procedure altlogregress(y : realarray; k : integer);
      (* Computes the Hurst exponent of a power spectrum y,
      indexed on 1,2,...,k, using data for 2,3,...,k-1.
      First, linear regression is used to fit log-
      transformed data to the equation ln(y) = a + b ln(i),
      yielding the slope  b  and the coefficient of
      correlation  r.  The Hurst exponent is given by  b
      = -1-2H, or H = -(b+1)/2. *)

var
   sumx : real;              (* sum of x-values *)
   sumy : real;              (* sum of y-values *)
   sumxsq : real;            (* sum of x**2      *)
   sumysq : real;            (* sum of xy        *)
   sumxy : real;             (* sum of y**2      *)
   a : real;                 (* intercept of
                             regression line *)
   b : real;                 (* slope of regression line *)
   sdb : real;               (* standard deviation of b *)
   r : real                  (* coefficient of correlation *)
   xbar,ybar : real;         (* average x, average y *)
   i : integer;

begin
    sumx := 0;
    sumy := 0;
    sumxy := 0;
    sumxsq := 0;
    sumysq := 0;


    for i := 2 to k-1 do
      begin
        sumx :=  sumx + ln(i);
        sumy :=  sumy + ln(y[i]);
        sumxy := sumxy + ln(i)*ln(y[i]);
        sumxsq := sumxsq + ln(i)*ln(i);
        sumysq := sumysq + ln(y[i])*ln(y[i]);
      end;

    k := k-2;                (* number of terms *)
    xbar := sumx/k ;
    ybar := sumy/k ;
```

```
   b := (sumxy - k*xbar*ybar)/(sumxsq - k*xbar*xbar);

   a := ybar - (b*xbar);

   r := (sumxy - k*xbar*ybar)/sqrt((sumxsq - k*xbar*xbar)
            *(sumysq - k*ybar*ybar));

   writeln('Hurst exponent    ', -(b+1)/2);
   if printflag = 'Y' then
      writeln(lst,'Hurst exponent     ', -(b+1)/2);

   if k > 4 then
      begin
         if ((r*r) < 1) then (* to avoid problems if r
           computes to 1+epsilon *)
            sdb := sqrt((1-r*r)*(sumysq-k*ybar*ybar)/
               ((k-4)*(sumxsq-k*xbar*xbar)))
         else
            sdb := 0;
         writeln('+/-  ',sdb/2);
         if k < 27 then
            writeln('Use t-test with ',k-2, ' degrees of
               freedom.');
      end
      else (* k <= 4 *)
         begin
               write;
               writeln('standard deviation undefined');
         end;
   writeln('coeff of correl.  ',r);

   if printflag = 'Y' then
      begin    (* output to printer *)
         if k > 4 then
           begin
             writeln(lst,'+/-  ',sdb/2);
             if k < 27 then
                writeln(lst,'Use t-test with ',k-2,
                  ' degrees of freedom.');
           end
           else (* k <= 4 *)
               begin
                   write(lst);
                   writeln(lst,'standard deviation
                      undefined');
               end;

         writeln(lst,'coeff of correl.  ',r);
      end;

end;  (* altlogregress *)

(*****************)
```

```
begin  (* main *)

    (* initialize *)
    for i := 1 to n do
       begin
          areal[i] := 0;
          aimag[i] := 0;
       end;

    writeln('Do you want output to printer (Y/N)');
    readln(printflag);
    if printflag = 'y' then
       printflag := 'Y';
    if printflag ='Y' then
       writeln('Please turn on the printer.');
    writeln('Type the name of the input data file.');
    writeln('Type "none" to enter data from keyboard');

    (* read in data *)
    xvalue := 1;    (* -9999 is flag for end of file *)

    readln(filename);
    if filename = 'none' (* data entered from keyboard *)
    then
        begin
            writeln('Enter the time series, one point at
              a time.');
            writeln('Follow each point with the "enter"
              key.');
            writeln('Number of points must be
              256.');
            writeln('Enter  -9999  to denote the end of
              data.');
            j := 1;
            repeat
                readln(xvalue);
                if xvalue > -9999 then
                   begin
                       areal[j] := xvalue;
                       j := j+1;
                   end;
                if j > n+1 then
                   begin
                        writeln('Too many points');
                        halt;
                   end;
            until xvalue < -9998;
            j := j-1;

            writeln('Length of time series = ',j);
            writeln('Input the maximum frequency.');
```

```
              writeln('At most 1/4 the length of the
                series.');
              readln(kk);
              if kk > 64 then
                 begin
                    writeln('Frequency too high');
                    halt;
                 end;
      end

else
      begin  (* data from input file *)
             if printflag ='Y' then
                writeln(lst,filename,'   ');
             assign(infile,filename);
             reset(infile);

             j := 1;
             repeat
                    readln(infile,xvalue);
                    if xvalue > -9999 then
                        begin
                           areal[j] := xvalue;
                           j := j+1;
                        end;
                    if j > n+1 then
                        begin
                              writeln('File too long');
                              close(infile);
                              halt;
                        end;
             until xvalue < -9998;
             j := j-1;
             close(infile);
             writeln('Length of file = ',j);
             writeln('Length must be 256.');
             writeln('Input the maximum frequency.');
             writeln('At most 1/4 the length of the
                file.');
             readln(kk);
             if kk > 64 then
                 begin
                    writeln('Frequency too high');
                    halt;
                 end;
      end; (* module for file input *)

if printflag ='Y' then
   writeln(lst,'maximum frequency = ',kk);
kk := kk+1; (* since fft is indexed on 1..n *)
if printflag ='y' then
   writeln(lst,'minimum frequency = 2');
```

```
writeln('minimum frequency = 2');
writeln('Shall we apply a median filter (Y/N) ?');
readln(filtflag);
if ((filtflag = 'Y') or (filtflag = 'y')) then
   if printflag = 'Y' then
      writeln(lst,'Median filter applied');

(* now the fft *)
initbitrev;
trigtable;
transform;

(* compute power spectrum, using symmetry of fft
- for fft of real data, argument of term at frequency
k  =  argument at frequency  n-k *)
for i := 1 to 256 do
   energy[i] := 0;
for i := 1 to kk do      (* i = frequency *)
   energy[i+1] := areal[i+1]*areal[i+1] + aimag[i+1]
         * aimag[i+1] + 0.000001;      (* to avoid
            underflow on log *)

(* median filter *)
if (filtflag = 'Y') or (filtflag = 'y') then
   for i := 2 to kk do      (* frequencies 2 to kk-1 *)
      begin                  (* bubble sort *)
         for j := 0 to 2 do
            median[j] := energy[i+j]; (* copy data at
               freq.i-1,i,i+1 since energy[k] is data
               for frequency  k-1 *)
         if median[0] > median[1] then
            begin
               temp := median[0]; (* interchange
                  median[0]and median[1],
                  making median[1] larger *)
               median[0] := median[1];
               median[1] := temp;
            end;
         if median[1] > median[2] then
            begin
               temp := median[1];
               median[1] := median[2];
               median[2] := temp;
            end;
         if median[0] > median[1] then
            begin
               temp := median[0];
               median[0] := median[1];
               median[1] := temp;
         energy[i] := median[1];          (* copy median
               to energy[i], frequency  i *)
      end
```

```
      else    (* no filter *)
        for i := 2 to kk do
             energy[i] := energy[i+1]; (* so energy[i]
                   now corresponds to frequency  i *)

    altlogregress(energy,kk);

end.
```

12.8 Other programs

We conclude with code for several other programs. The program *pwrtest1*
below generates the spectrum of a Mandelbrot–Weierstrass fractal except
that the amplitudes are not random. Instead, the amplitudes scale exactly
as $c/n^{0.5+H}$. This program can be used to generate test data for the power
spectrum program above.

12.8.1 *Program for test data for power spectrum (pwrtest1)*

```
program pwrtest1;
      (* Derived from fftmwfract.  Generates and graphs
      Mandelbrot- Weierstrass fractals in one dimension
      using (inverse) fast Fourier transform applied to a
      power spectrum with amplitude exactly  freq**-(H+0.5)
      and random phases.  *)

      (* Writes data to a file called testdata.dat for
      testing other programs *)

uses
    crt, graph3;

var
    y : array [0..256] of real;  (* coordinate on screen *)
    i,j : integer;
    h : real;                  (* Hurst exponent *)
    hplus : real;              (* Hurst exponent plus 1/2 *)
    terms : integer;           (* highest frequency in
                                 spectrum, at most 127 *)
    areal,aimag : array[1..256] of real; (* real and imag
                                 parts of spectrum *)
    ymax : real;               (* maximum y to scale image *)
    scale : real;              (* to scale image *)
    temp : real;               (* temporary real *)
    twopidiv : real;           (* 2*pi / 256, so as to graph
                                 one complete cycle in the 256
                                 pixel width used *)
```

```
      index : array [1..256] of integer;   (* for bit reversal
                        in fft *)
      costable, sintable : array [0..256] of real;   (* for
                        fft *)

      outfile : text;
      flag : string[2];                    (* for caution *)

const
    n = 256;            (* number of terms, size of fft *)
    hn = 128;           (* half n *)
    log2n = 8;          (* log to base 2 of n *)


procedure initbitrev; (* constructs index array for bit
    reversal in fft *)
    var k,l,lmax,ii,m : integer;

    begin
       index[1] := 1;
       lmax := 1;
       ii := hn;

       for k := 1 to log2n do
          begin
             for l := 1 to lmax do
                index[l+lmax] := index[l] + ii;
                ii := ii div 2;
                lmax := 2 * lmax;
             end;
    end;   (* procedure initbitrev *)

procedure trigtable;
    var
       k : integer;
       angle : real;

    begin
       for k := 0 to n do
          begin
             angle := 2*pi*k/n;
             costable[k] := cos(angle);
             sintable[k] := sin(angle);
          end;
    end;   (* procedure trigtable *)

procedure transform;
    var
       i,i0,i1,j,l,p,hp, ndivp : integer;
       tempreal, tempimag : array[1..256] of real;
       c,s : real;   (* cosine, sine *)
       t1, t2 : real;   (* temporaries *)
```

```
begin
    (* first do bit reversal *)
    for i := 1 to n do
        begin
            i0 := index[i];
            tempreal[i] := areal[i0];
            tempimag[i] := aimag[i0];
        end;
        for i := 1 to n do
            begin
                areal[i] := tempreal[i];
                aimag[i] := tempimag[i];
            end;

        (* initialize some variables *)
        p := 2;
        hp := 1;
        ndivp := n div p;

        for l := 1 to log2n do
            begin
                for j := 1 to hp do
                    begin
                        c := costable[(j-1)*ndivp];
                        s := -sintable[(j-1)*ndivp];
                        (* inverse FFT *)
                        i := j;
                        for i1 := 1 to ndivp do
                            begin
                                i0 := i + hp;
                                t1 := areal[i0]*c - aimag[i0]*s;
                                t2 := areal[i0]*s + aimag[i0]*c;
                                areal[i0] := areal[i] - t1;
                                aimag[i0] := aimag[i] - t2;
                                areal[i] := areal[i] + t1;
                                aimag[i] := aimag[i] + t2;
                                i := i + p;
                            end;  (* for i1 *)
                    end;  (* for j *)
                hp := p;
                p := 2*p;
                ndivp := ndivp div 2;
            end;  (* for l *)

end;  (* procedure transform *)

begin  (* main *)

    (* caution *)
    writeln('CAUTION - WILL OVERWRITE testdata.dat.
      CONTINUE (Y/N) ?');
    readln(flag);
    if ((flag <> 'Y') and (flag <> 'y')) then
```

```
   halt;

(* else continue *)

(* initialize *)
twopidiv := 2*pi/256;   (* Turbo Pascal has pi as a
   predefined const. *)
writeln('What is the Hurst exponent H (0 to 1 (rough to
   smooth)) ?');
readln(h);
hplus := h + 0.5; (* to simplify calculating the
   spectrum *)
writeln('Highest frequency (at most 127) ?');
readln(terms);
writeln('Program will graph fractal.');
writeln('Type "enter" to clear screen and write data to
   file when done.');


(* setups for FFT *)
initbitrev;
trigtable;

for i := 1 to n do
   begin
      areal[i] := 0;
      aimag[i] := 0;
   end;   (* for i *)

(* generate the spectrum *)
for i := 2 to terms+1 do
   begin
      temp := 2*pi*random; (* for random phases *)
      areal[i] := cos(temp)*exp(-hplus*ln(i-1));
      aimag[i] := sin(temp)*exp(-hplus*ln(i-1));
   end;   (* for i *)

   (* generate y-values by applying fast Fourier
   transform -- the inverse FFT is the conjugate of the
   FFT multiplied by a scale factor, and we only need
   the real part of the inverse FFT *)

transform;
ymax := 0;   (* will hold maximum absolute value of y
   for scaling *)
for i := 1 to n do (* loop over x-values, in pixels *)
if abs(areal[i]) > ymax
   then ymax := abs(areal[i]);
for i := 1 to n do
   y[i] := 90*areal[i]/ymax;

(* plot the fractal *)
graphcolormode;
```

```
    y[0] := y[n];
    for i := 0 to n-1 do
        draw (i,100-round(y[i]),i+1,100-round(y[i+1]),3);
        (* line from ith to (i+1)th point *)

    readln;
    textmode(80);
    assign(outfile,'testdata.dat');
    rewrite(outfile);
    for i := 1 to 256 do
        writeln(outfile,y[i]);
    writeln(outfile,-9999);   (* flag for eof *)
    writeln(outfile,h);
    close(outfile);
end.
```

The last program, a simple iterated function system (Barnsley *et al.* 1986, Barnsley 1988), was used to generate a finite random subset of the Sierpiński triangle in Section 8.3 above. This program, called *fractgame*, generates a sequence of points $\{x(t)\}$ as follows. Fix n vertices of a polygon $p_1, p_2, \ldots, p_n$, and a parameter r $(0 < r < 1)$. Our program requires $n \leqslant 16$. At each time step t, one vertex p_j is chosen randomly, and the next point $x(t+1)$ is chosen along the line from $x(t)$ to the selected vertex p_j according to

$$x(t+1) = x(t) + rp_j. \tag{12.5}$$

If $n = 3$ and $r = 0.5$, and the vertices p_1, p_2, p_3 form an equilateral triangle, then the above process generates a Poisson distribution on the Sierpiński triangle (see Barnsley 1988). Recall that the one-dimensional game of life (Section 7.2) also generated a Sierpiński triangle (Fig. 7.1, left). The reader can easily show that if r is sufficiently large then the process generates a fractal of dimension

$$D = -\log n/\log(1 - r). \tag{12.6}$$

12.8.2 *Program for fractal game (fractgame)*

```
program fractgame;
        (*  An iterated function system (Barnsley et al.
        1986, Barnsley 1988) which generates a random
        distribution on a fractal.  The system evolves as
        follows.  One of several vertices is selected at
        random.  Then the current point (state) is replaced
        by a weighted average of the current point and the
        selected vertex.  *)
uses
        crt, graph3;
```

```
var
    i,j,k: integer;
    pts: integer;      (*  number of points *)
    p,q: array [0..15] of integer; (* coords of vertices *)
    x,y: real;         (* coords of present point *)
    rat: real          (* fraction of distance to move *)
    timesteps: integer;

begin
    writeln('A simple iterated function system; see
            Barnsley et al. 1986, Barnsley, 1988.');
    writeln('Program will draw fractal.  At completion,
            type');
    writeln('"enter" to clear screen.');
    writeln('How many points ? At most 16.  Enter 3 to
            generate');
    writeln('the Sierpinski triangle');
    readln(pts);
    writeln('Enter the coords of the points, in the form
            100 200');
    for i := 0 to pts-1 do
        readln(p[i],q[i]);
    writeln('Enter the starting point');
    readln(x,y);
    writeln('Enter fraction of distance (0 to 1) to move
            toward');
    writeln('selected vertex,');
    writeln('0.5 for the Sierpinski triangle');
    readln(rat);
    writeln('Enter the number of time steps.  Enter about
            1000 to');
    writeln('largely fill the Sierpinski triangle, about
            20 for a');
    writeln('random distribution on the triangle.');
    readln(timesteps);
    writeln('Program will draw fractal.  Type "enter" to
            clear screen');
    writeln('when done.  Type "enter" to begin');
    readln;
    graphcolormode;
    for i := 1 to timesteps do
        begin
            j := trunc(pts*random);
            x := x + rat*(p[j]-x);
            y := y + rat*(q[j]-y);
            plot(trunc(x),trunc(y),3);
        end;
    readln;
    textmode(80);
end.
```

Annotated references

We include reactions and eclectic comments to some of the references below. Citations in **boldface** are especially recommended for further reading.

Adler, R. J. (1981). *The geometry of random fields.* Wiley, New York.

Aho, A., Hopcroft, J. E., and Ullman, J. D. (1974). *Design and analysis of computer algorithms.* Addison-Wesley, Reading, MA, USA.

Allen, T. F. H. and Starr, T. B. (1982). *Hierarchy: Perspectives for ecological complexity.* University of Chicago Press.

Bak, P., Tang, C., and Weisenfeld, K. (1987). Self-organized criticality: An explanation of $1/f$ noise. *Phys. Rev. Lett.* **59**, 381–84.

Barnsley, M. F. (1988). *Fractals everywhere.* Academic Press, Boston.

Barnsley, M. F., Ervin, V., Hardin, D., and Lancaster, J. (1986). Solution of an inverse problem for fractals and other sets. *Proc. Natl. Acad. Sci. USA* **83**, 1975–7.

Berry, M. V. and Lewis, C. V. (1980). On the Weierstrass–Mandelbrot fractal function. *Proc. R. Soc. Lond.* A **370**, 459–84.

Billingsley, P. (1967). *Ergodic theory and information.* Wiley, New York.

Bradbury, R. H., Reichlet, R. E., and Green, D. G. (1984). Fractals in ecology: methods and interpretation. *Marine Ecol. Prog. Ser.* **14**, 295–6.

Burrus, C. S. and Parks, T. W. (1985). *DFT/FFT and convolution algorithms.* Wiley, New York.

Cantor, G. (1872). Über die Ausdehnung eines Satzes aus der Theorie der Trigonometrischen Reihen. *Math. Annalen* **5**, 123–32. *The construction of the Cantor set.*

Carathéodory, C. (1914). Über das lineare Maß von Punktmengen—eine Verallgemeinerung des Längebegriffs. *Nachrichten der K. Gesellschaft der Wissenschaften zu Göttingen. Mathematisch-physikalische Klasse,* 404–26. Reprinted in *Gessamelte mathematisches Schriften,* Beck, Munich, vol. 4, 1956, pp. 249–75. *Forerunner of Hausdorff (1919) measure.*

Carlson, J. M. and Langer, L. S. 1989. Properties of earthquakes generated by fault dynamics. *Phys. Rev. Lett.* **62**, 2632–5.

Caserta, F., Stanley, H. E., Eldred, W. D., Daccord, G., Hausman, R. E., and Nittman, J. (1990). Physical mechanisms underlying neurite outgrowth: A quantitative analysis of neuronal shape. *Phys. Rev. Lett.* **64**, 95–9.

Collet, P. and Eckmann, J.-P. (1980). *Iterated maps of the interval as dynamical systems,* Progress in Physics, Vol. 1. Birkhäuser, Boston.

Cooley, J. M., Lewis, P. A., and Welch, P. D. (1967). History of the fast Fourier transform. *Proc. IEEE* **55**, 1675–7.

Cooley, J. M. and Tukey, J. W. (1965). An algorithm for machine calculation of complex Fourier series. *Math. Comp.* **19**, 297–301.

Danielson, C. C. and Lanczos, C. (1942). Some improvements in practical Fourier analysis and their applications to X-ray scattering from liquids. *J. Franklin Inst.* **233**, 365–80, 435–52.

Draper, N. and Smith, H. (1981). *Applied regression analysis*, 2nd edn. Wiley, New York.

du Bois, R. P. (1875). Versuch einer Klassification der willkürlichen Functionen reeler Argument nach ihren Änderungen in den kleinsten Intervallen. *J. für die reine und angewandte Mathematik* **79**, 21–37. *Weierstrass's continuous nondifferentiable function—a mathematical monster of the 1870s.*

Dugundji, J. (1966). *Topology.* Allyn and Bacon, Boston.

Ehrlich, P. R. (1965). The population biology of the butterfly *Euphydryas editha*. II. The structure of the Jasper Ridge Colony. *Evolution*, **19**, 327–36.

Ehrlich, P. R. and Murphy, D. D. (1981). The population biology of checkerspot butterflies (*Euphydryas*)—a review. *Biologisches Zentralblatt*, **100**, 613–29.

Erzan, A. and Sinha, S. (1991). Spatiotemporal intermittency of the sandpile. *Phys. Rev. Lett.* **66**, 2750–3.

Falkmer, S. (1985). Comparative morphology of pancreatic islets in animals. In *The diabetic pancreas* (ed. Volk, B. W. and Arquilla, E. R.), Plenum Press, New York, pp. 17–52.

Fleischmann, M., Tildesley, D. J., and Ball, R. C. (eds.) (1989). *Fractals in the natural sciences.* Princeton University Press.

Frontier, S. A. (1987). Fractals in marine evology. In *Developments in numerical ecology* (ed. Legendre, L.), Springer-Verlag, Berlin, pp. 335–78.

Gardner, M. (1983). *Wheels, life, and other mathematical amusements.* Freeman, New York. *This is perhaps the best reference to Conway's game of life.*

Ghil, M., Kimoto, M., and Neelin, J. D. (1991). Nonlinear dynamics and predictability in the atmospheric sciences. *Reviews of Geophysics, Supplement, April 1991, US National Report of the International Union of Geodesy and Geophysics*, pp. 46–55. *This paper provides a good review of the problems of identification and predictability in complex nonlinear systems.*

Good, I. J. (1958). The interaction algorithm and practical Fourier series. *J. Royal Statist. Soc. Ser. B* **20**, 361–72. Addendum, **22** (1960), 372–5.

Gorray, K. C., Baskin, D. G., and Fujimoto, W. Y. (1986). Physiological and morphological changes in islet B cells following alloxan treatment. *Diabetes Res.,* **3**, 187–91.

Gorray, K. C., Baskin, D. G., Brodsky, J., and Fujimoto, W. Y. (1986). Responses of pancreatic B cells to alloxan and streptozotocin in the guinea pig. *Pancreas*, **1**, 130–8.

Grassberger, P. and Procaccia, I. (1983). Characterization of strange attractors. *Phys. Rev. Lett.* **50**, 346–9.

Gutenberg, B. and Richter, C. F. (1942). Earthquake magnitude, intensity, energy, and acceleration. *Bull. Seismol. Soc. Am.* **32**, 163–91. *See Gutenberg and Richter (1956b).*

Gutenberg, B. and Richter, C. F. (1954). *Seismicity of the earth.* Princeton University Press.

Gutenberg, B. and Richter, C. F. (1956a). Earthquake magnitude, intensity, energy, and acceleration. *Bull. Seismol. Soc. Am.* **46**, 105–145. *See Gutenberg and Richter (1956b).*

Gutenberg, B. and Richter, C. F. (156b). Magnitude and energy of earthquakes. *Annali di Geofisica (Roma)* **9**, 1–15.

Harrison, S., Quinn, J. F., Baughman, J. F., Murphy, D. D., and Ehrlich, P. R. (1986). The effects of scientific study on two butterfly populations. *Am. Natur.* **137**, 227–43.

Harrison, S., Quinn, J. F., Baughman, J. F., Murphy, D. D., and Ehrlich, P. R. (1991). Estimating the effects of scientific study on two butterfly populations. *Amer. Natur.*, **137**, 227–43.

Hastings, H. M. (1983). Stability of community interaction matrices. *Current Trends in Food Web Theory* (ed. D. DeAngelis, M. Post, and G. Sugihara) Oak Ridge National Laboratory Technical Report, ORNL-5983 pp. 65–8.

Hastings, H. M., Pekelney, R., Monticciolo, R., vun Kannon, D., and del Monte, D. (1982). Time scales, persistence and patchiness. *BioSystems* **15**, 281–9.

Hastings, H. M., Schneider, B. S., Schreiber, M., Gorray, K., Maytal, G., and Maimon, J. (1992). Statistical geometry of pancreatic islets, *Proc. Royal. Soc. Lond. Ser. B.* **250**, 257–61.

Hastings, H. M. and Troyan, D. T. (1991). Fractal analysis of earthquake time series. Abstract, AGU Fall 1991 Meeting Abstracts, Eos, supplement, 29 October, p. 291. *See Section 8.2.*

Haury, L. R., McGowan, J. A., and Wiebe, P. H. (1978). Patterns and processes in the time-space scales of plankton distributions. In: Steele, J. H. (ed.) (1978), pp. 277–327. *A nice exposition of the Stommel diagram which summarizes observed scales of variability in marine processes.*

Hausdorff, F. (1919). Dimension und äusseres Mass. *Mathematisches Annalen* **79**, 157–79.

Hellerstrom, C. and Swenne, I. (1985). Growth patterns of pancreatic islets in animals. In *The diabetic pancreas* (ed. Volk, B. W. and Arquilla, E. R.), Plenum Press, New York, pp. 53–79.

Hentschel, H. G. E. and Procaccia, I. (1983). The infinite number of generalized dimensions of fractals and strange attractors. *Physica* **8D**, 435–44.

Hirsch, M. W. and Smale, S. (1974). *Differential equations, dynamical systems, and linear algebra.* Academic Press, New York.

Hogg, R. V. and Tanis, E. A. (1977). *Probability and statistical inference.* Macmillan, New York.

Holmes, R. T., Sherry, T. W., and Sturges, F. W. (1986). Bird community dynamics in a temperate deciduous forest: Long-term trends at Hubbard Brook [New Hampshire, USA]. *Ecol. Monogr.* **56**, 201–20.

Horowitz, E. and Sahni, S. (1984). *Fundamentals of computer algorithms.* Computer Science Press, Rockville, MD, USA.

Hurewicz, W. and Wallman, H. (1941). *Dimension theory.* Princeton University Press. *A classic.*

Hurst, H. E. (1951). Long-term storage capacity of reservoirs. *Trans. Am. Soc. Civil Eng.* **116**, 770–808. *Hurst found that fluctuations in Nile River outflows display long-term correlations. See Mandelbrot (1977, 1982) for a modern reference.*

Hurst, H. E. (1956). Methods of using long-term storage in reservoirs. *Proc. Inst. Civil Eng.* **5**, Part 1, 519–77.

Hutchinson, G. E. and MacArthur, R. H. (1959). A theoretical ecological model of size distributions among species of animals. *Am. Natur.* **93**, 117–25.

Jackson, J. B. and Hughes, T. (1985). Adaptive strategies of coral reef invertebrates. *Am. Sci.* **73**, 265–74.

Johnston, J. (1973). *Econometric methods*, 2nd edn. McGraw-Hill, New York. *A practical reference for applied statistics.*

Kagan, Y. Y. and Jackson, D. D. (1991). Seismic gap hypothesis: Ten years after. *J. Geophys. Res.* **96**, 21, 419–31. *Are you safer just after an earthquake. See Section 8.2. Monastersky (1992) provides a nice review of this paper.*

Kleinfeld, D., Raccuia-Behling, F., and Blonder, G. E. (1990). Comments on physical mechanisms underlying neurite outgrowth: A quantitative analysis of neuronal shape. *Phys. Rev. Lett.* **64**, 3064.

Knuth, D. E. (1981). *The art of computer programming*, Vol. 2: *Seminumerical algorithms.* Addison-Wesley, Reading, MA, USA. *The standard reference in the theory behind applied computer science; includes a good survey of random number generators.*

Koch, H. von (1906). Une méthode géométrique élémentaire pour l'étude de certaines questions de la théorie des courbes plaines. *Acta Mathematica* **30**, 145–74. *The Koch snowflake, a continuous nowhere-differentiable curve, which has become a standard example in the study of fractals. See Mandelbrot (1977, 1982).*

Korcak, J. (1938). Deux types fondamentaux de distribution statistique. *Bull. de l'Institut International de Statistique*, III, 295–9.

Kreyszig, J. (1988). *Advanced engineering mathematics*, 6th edn. Wiley, New York.

Krummel, J. R., Gardner, R. H., Sugihara, G., and O'Neill, R. V. (1987). Landscape patterns in a disturbed environment. *Oikos* **48**, 321–84.

LeNormand, R. (1989). Flow through porous media: Limits of fractal patterns. In: Fleischmann, M., Tildesley, D. J., and Ball, R. C. (eds.) (1989), pp. 159–67.

Levin, S. A. and Paine, R. T. (1974). Disturbance, patch formation and community structure. *Proc. Natl. Acad. Sci. USA* **71**, 2744–7.

Li, T. and Yorke, J. A. (1975). Period three implies chaos. *Am. Math. Monthly* **82**, 985–92.

Lin, C. C. and Segel, L. A. (1974). *Mathematics applied to deterministic problems in the natural sciences.* Macmillan, New York. *A good introductory graduate text, cited here for its treatment of random walks, Brownian motion, and diffusion.*

Linsker, R. (1988). Self-organization in a perceptual network. *IEEE Computer*, March, pp. 105–17.

Logothetopolis, J. (1972). Islet cell regeneration and neogenesis. In *Endocrine Pancreas Handbook of Physiology* Vol. 1 (ed. Steiner, D. F. and Freinkel, N.), Wilkinson and Wilkins Co., Baltimore, MD, pp. 67–76.

Lorenz, E. N. (1963). Deterministic nonperiodic flow. *J. Atmos. Sci.*, **26**, 636–46.

Lovejoy, S. (1982). Area–perimeter relationship for rain and cloud areas. *Science* **216**, 185–7.

Lovejoy, S., and Schertzer, D. (1991). Multifractal analysis techniques and the rain and cloud fields from 10^{-3} to 10^6 m. In: Schertzer, D. and Lovejoy, S. (eds.) (1991), pp. 111–44.

Lovejoy, S., Schertzer, D., and Ladoy, P. 1986. Fractal characterization of inhomogeneous geophysical measuring networks. *Nature* **319**, 43–4.

Luo, X. and Schramm, D. N. (1992). Fractals and cosmological large-scale structure. *Science* **256**, 513–15.

MacArthur, R. H. and Wilson, E. O. (1967). *The theory of island biogeography.* Princeton University Press.

McCaffrey and Hamilton (1978). *Vegetation map of the Okefenokee Swamp.* University of Georgia. Athens, GA, USA.

McGuire, M. (1991). *An eye for fractals.* Addison-Wesley, Redwood City, CA, USA. *Beautiful photographs and good intuitive explanations. A nice introduction to fractal dimension* (see our Chapter 3).

Mandelbrot, B. B. (1965). Une classe de processus stochastiques homothetiques a soi; application a la loi climatologique de H. E. Hurst. *Comptes Rendus Acad Sci. Paris* **260**, 3274–7.

Mandelbrot, B. B. (1975). *Les objects fractals: forme, hasard et dimension.* Flammarion, Paris. *Introduces the term* fractal.

Mandelbrot, B. B. (1977). *Fractals: Form, chance and dimension.* Freeman, San Francisco. *English translation of Mandelbrot (1975).*

Mandelbrot, B. B. (1982). *The fractal geometry of nature.* Freeman, San Francisco.

Mandelbrot, B. B. (1989). Fractal geometry: What is it and what does it do? In: Fleischmann, M., Tildesley, D. J. and Ball, R. C. (eds.) (1989), pp. 3–16.

Mandelbrot, B. B. and Van Ness, J. W. (1968). Fractional Brownian motions, fractional noises and applications. *SIAM Rev.* **10**, 422–37.

Mandelbrot, B. B. and Wallis, J. R. (1969). Some long-run properties of geophysical records. *Water Resources Res.* **5**, 321–40.

Marsaglia, G. (1968). Random numbers fall mainly in the planes. *Proc. Natl. Acad. Sci. USA* **61**, 25–8.

May, R. M. (1972). Will a large complex system be stable? *Nature,* **238**, 413–14.

May, R. M. (1974). *Stability and complexity in model ecosystems.* Princeton University Press.

May, R. M. (1975). Patterns of species abundance and diversity. In: *Ecology and evolution of communities* (ed. M. L. Cody and J. M. Diamond). Harvard University Press.

May, R. M. (1978). The dynamics and diversity of insect faunas. In: *Diversity of insect faunas* (ed. I. A. Mound and N. Waloff). Blackwell, Oxford, pp. 188–204.

May, R. M. (1988). How many species are there on earth? *Science* **241**, 1441–8.

May, R. M. (1990). How many species? *Phil. Trans. Royal Soc. Lond.* B **330**, 171–81.

Meakin, P. and Tolman, S. (1989). Diffusion-limited aggregation. In: (Fleischmann, M., Tildesley, D. J., and Ball, R. C. (eds.) (1989). Pp. 133–47.

Meltzer, M. I. (1990). Livestock biotechnology: the economic and ecological impact of alternatives for controlling ticks and tick-borne diseases in Africa. Ph.D. thesis, Cornell University, Ithaca, NY, USA.

Meltzer, M. I. and Hastings, H. M. (1992). The use of fractals to assess the ecological impact of increased cattle population: case study from the Runde Communal Land, Zimbabwe. *J. Appl. Ecol.,* **29**, 635–46.

Milne, B. T. (1988). Measuring the fractal dimension of landscapes, *Appl. Math and Computation,* **27**, 67–79.

Monastersky, R. (1992). Shaking up seismic theory, *Science News,* **141**, 136–7.

Morse, D. R., Lawton, J. H., Dodson, M. M., and Williamson, M. H. (1985). Fractal dimension of vegetation and the distribution on anthropod body lengths. *Nature* **314**, 731–3.

Murphy, D. D., Menninger, M. S., Ehrlich, P. R., and Wilcox, B. A. (1986). Local population dynamics of adult butterflies and the conservation status of two closely related species, *Biol. Conserv.* **37**, 201–24. *Time series of* Euphydryas *populations used in Chapter 11.*

Odum, E. P. (1953). *Fundamentals of ecology*, Saunders, Philadelphia.

O'Neill, R. V., DeAngelis, D. L., Waide, J. B., and Allen, T. F. H. (1986). *A hierarchical concept of ecosystems*. Princeton University Press.

Paumgartner, D., Losa, G., and Weibel, E. R. (1981). Resolution effects upon the stereological estimation of surface and volume and its interpretation in terms of fractal dimensions. *J. Microsc.* **121**, 51–63.

Perrin, J. (1906). La discontinuité de la matière. *Revue de Mois* **1**, 323–44.

Peters, R. H. (1983). *The ecological implications of body size.* Cambridge University Press, *Morse* et al. *(1985) use these results for a fractal interpretation of habitat on leaves.*

Press, W. H., Vlannery, B. P., Teukolsky, S. A., and Vetterling, W. T. (1989). *Numerical recipes in Pascal.* Cambridge University Press. *Also available in C and Fortran, and software also available on disk. The standard handbook for scientific computing.*

Preston, F. W. (1962). The canonical distribution of commonness and rarity. *Ecology* **43**, 185–215, 410–432.

Rasmussen, E. M., Wang, X., and Ropelewski, C. F. (1990). The biennial component of ENSO variability. *J. Marine Syst.* **1**, 71–96.

Richardson, L. F. (1961). The problem of contiguity: An appendix of statistics of deadly quarrels. *General Systems Yearbook* **6**, 139–87. *Richardson considers the problem of measuring the length of a coastline. Mandelbrot (1977, 1982) cites and explains his work, which is now called the dividers method.*

Ripley, B. D. (1981). *Spatial statistics.* Wiley, New York.

Rogers, C. A. (1970). *Hausdorff dimension.* Cambridge University Press.

Ruelle, D. (1990). Deterministic chaos: The science and the fiction. *Proc. R. Soc. Lond.* A **437**, 241–8.

Runge, C. and König, H. (1924). *Die Grundlehren der mathematischen Wissenschaften*, Vol. 11, Springer, Berlin.

Schertzer, D. and Lovejoy, S. (1991). *Non-linear variability in geophysics: Scaling and fractals.* Kluwer, Dordrecht, The Netherlands.

Scheuring, I. (1991). The fractal nature of vegetation and the species–area relation. *Theor. Pop. Biol.* **39**, 170–7.

Schlesinger, W. H. (1978). Community structure, dynamics and nutrient cycling in the Okefenokee cypress swamp-forest. *Ecol. Monogr.* **48**, 43–66.

Schmidt, R. F. (ed.) (1978). *Fundamentals of neurophysics.* Springer, New York.

Schmidt-Nielsen, K. (1984). *Scaling: Why is animal size so important?* Cambridge University Press. *See comment under Peters (1983).*

Short, N. M., Lowman Jr, P. D., Freden, S. C., and Finch Jr, W. A. (1976). *Mission to Earth: Landsat views the world.* National Aeronautics and Space Administration, US Government Printing Office, Washington, DC.

Smale, S. and Williams, R. F. (1976). The qualitative analysis of a differential equation of population growth. *J. Math. Biol.* **3**, 1–4.

Smith, L. A. (1988). Intrinsic limits on dimension calculations. *Phys. Lett.* A **113**, 283–8.

Smoot, G. F., Bennett, C. L., Kogut, A., Wright, E. L., Aymon, J., Boggess, N. W., Cheng, E. S., DeAmici, G., Gulkis, S., Hauser, M. G., Hinshaw, G., Lineweaver, C., Loewenstein, K., Jackson, P. D., Jannsen, M., Kaita, E., Kelsall, T., Keegstra, P., Lubin, P., Mather, J., Meyer, S. S., Moseley, S. H., Murdock, T., Rooke, L., Silverberg, R. F., Tenorio, L., Weiss, R., and Wilkinson, D. T. (1992). Structure in the COBE DMR first year maps. *Astrophys. J. Lett.* **396**, L1. *The very early universe was not isotropic at a resolution of 1 in 10^{-6}.*

Steele, J. H. (ed.) (1978). *Spatial pattern in plankton communities*, Proc. NATO Conf. on Marine Biology. Plenum, New York.

Stommel, H. (1963). Variaties of oceanographic experience. *Science* **139**, 572–6.

Stommel, H. (1965). Some thoughts on planning the Kuroshio survey. In: *Proc. Symp. on the Kuroshio*, Tokyo, 1963, Oceanographic Society of Japan and Unesco. *See Haury* et al. *(1978)*.

Sugihara, G. (1980). Minimal community structure: An explanation of species abundance patterns. *Am. Natur.* **116**, 770–87.

Sugihara, G. (1982). Applications of fractals to the study of landscape patterns, unpublished. Oak Ridge National Laboratory technical memo.

Sugihara, G. (rapporteur) (1984). Ecosystem dynamics. In *Dahlem Konferenzen Exploitation of Marine Communities* (ed. R. M. May). Springer, Berlin, pp. 131–53.

Sugihara, G. and May, R. M. (1990*a*). Applications of fractals in ecology. *Trends Ecol. Evol.* **5**, 79–86.

Sugihara, G. and May, R. M. (1990*b*). Nonlinear forecasting as a way of distinguishing chaos from measurement error in time series. *Nature* **344**, 734–41.

Szalay, A. S. and Schramm, D. N. (1985). Are galaxies more closely correlated than clusters? *Nature* **314**, 718–20. *Is the distribution of stars fractal, and, if so, why? Luo and Schramm (1992) suggest an answer; see also Smoot* et al. *(1992) for experimental data.*

Tang, C. and Bak, P. (1988). Critical exponents and scaling relations for self-organized critical phenomena. *Phys. Rev. Lett.* **60**, 2347–50.

Tanner, J. T. (1966). Effects of population density on growth rates of animal populations. *Ecology* **47**, 733–45.

Tilton, J. C. (1987). Contextual classification on the massively parallel processor. In: *Frontiers of massively parallel scientific computation* (ed. J. R. Fischer), NASA Conf. Publication 2478. NASA, Greenbelt, MD, pp. 171–81.

Tont, S. A. (1975). The effect of upwelling on solar irradiance near the coast of southern California. *J. Geophys. Res.* **80**, 5031–4.

Tont, S. A. (1981). Upwelling: effects on air temperature and solar irradiance. *Coastal Upwelling and Estuarine Studies* **1**, 57–62.

United States Geological Survey (1973). Map of the Natchez Quadrangle. USGS, Reston, VA.

Utsu, T. (1988). A catalog of large earthquakes ($M \geq 6$) and damaging earthquakes in Japan for the years 1885–1925. In: *Historical seismographs of earthquakes of the world* (ed. W. H. K. Lee, H. Meyers, and K. Shimazaki). Academic Press, San Diego, pp. 150–61.

Vicsek, T. (1989). *Fractal growth phenomena*. World Scientific, Riveredge, NJ, USA.

Weibel, E. R. (1979). *Stereological methods. Vol. 1 Practical methods for biological morphometry*. Academic Press, New York.

Wickman, B. and Hill, D. (1987). Building a random-number generator. *Byte*, March, pp. 127–9.

Witten, T. A. and Sander, L. M. (1981). Diffusion-limited aggregation, a kinetic critical phenomenon. *Phys. Rev. Lett.* **47**, 1400–1403. *Models for fractal growth; see our Sections 8.3 and 8.4 for applications to developmental biology.*

Wolfram, S. (1984). Cellular automaton models as models of complexity. *Nature* **311**, 419–24.

Wolfram, S. (1985). Origins of randomness in physical systems. *Phys. Rev. Lett.* **55**, 449–53. *Randomness can arise in many simple nonlinear systems, including simple cellular automata models. See Wolfram (1984) for a nice introduction.*

Wolfram, S. (ed.) (1986). *Theory and applications of cellular automata* World Scientific, Singapore.

Index

Fractals: A User's Guide for the Natural Sciences explains Mandelbrot's fractal geometry and describes some of its applications in the natural world. Written to enable students and researchers to master the methods of this topical subject, the book steers a middle course between the formality of many papers in mathematics and the informality of picture-oriented books on fractals. It is both a logically developed text and a 'fractals for users' handbook.

Fractal geometry exploits a characteristic property of the real world—self-similarity—to find simple rules for the assembly of complex natural objects. Beginning with the foundations of measurement in Euclidean geometry, the authors progress from analogues in the geometry of random fractals to illustrative applications spanning the natural sciences: the developmental biology of neurons and pancreatic islets; fluctuations of bird populations; patterns in vegetative ecosystems; and even earthquake models. The final section provides a toolbox of user-ready programs. This volume is an essential resource for all natural scientists interested in working with fractals.

ISBN 0-19-854597-5

9 780198 545972

OXFORD UNIVERSITY PRESS